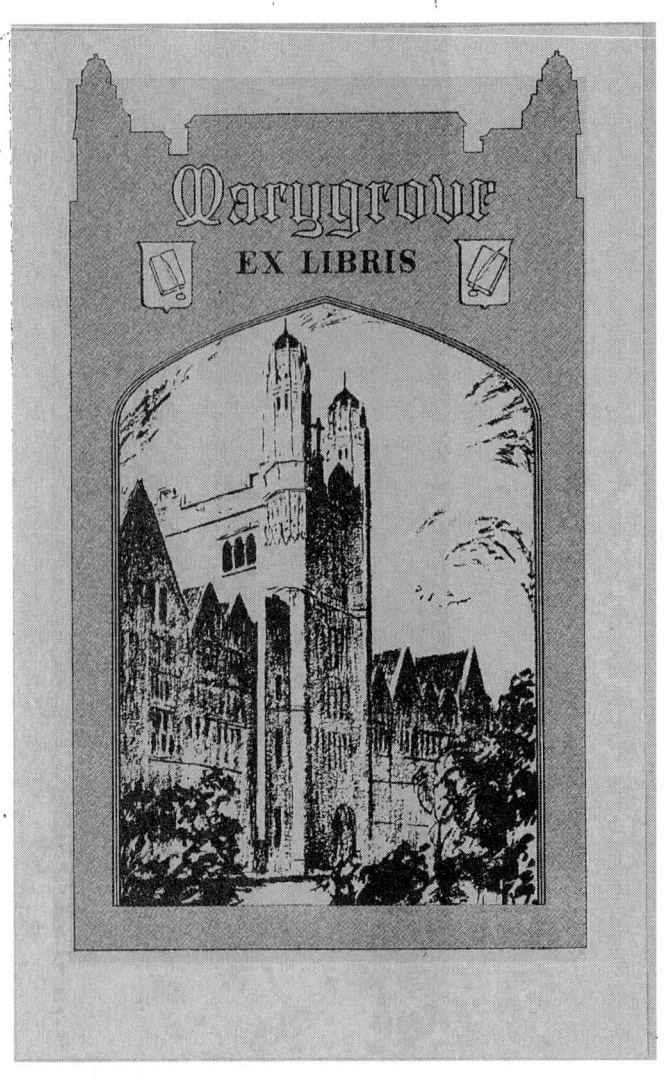

940.5418
G76

WAR IS MY PARISH

War Is My Parish

Anecdote and Comment
Collected by
DOROTHY FREMONT GRANT

THE BRUCE PUBLISHING COMPANY
Milwaukee, Wisconsin

To

The Chaplains of World War II

Who Have Died "In Line of Duty"

REQUIESCANT IN PACE

V-MAIL...JULY, 1944

Manhasset, Long Island, N. Y.
Feast of St. Ignatius of Loyola

Dear Fellow-Catholic:

Why have you chosen to read this book? Is your son, your brother, your cousin or perhaps your uncle a Catholic Chaplain? Do you scan these pages to seek his name only? If so, an explanation is due you at the start, for herein you may not find the name of your Chaplain kin.

Before undertaking this project I confided it to an eminent prelate. He looked at me kindly, but gravely shook his head. "It will be dynamite, my child — dynamite. You cannot possibly make it comprehensive and all inclusive; a properly revealing document."

I agreed this would be impossible. I am told that ten percent of the priests in the United States are now in service. Of course, I could not include all of this ten percent in these pages: shall I then refuse to compile them? Shall I sit back and say nothing? It has always been hard for me to keep still when something needed saying.

So, I replied to the eminent prelate, "I do not aim to write a 'document.' That is the task of scholars. But why is there dynamite in trying to say something that needs to be said?"

"For example?" my priestly friend urged me on.

"For example," I took him up, "our Chaplains deserve a 'thank you' for what they are doing; our country deserves a 'thank you' for having the good sense to send

Chaplains off with our men; — God should be thanked for the Chaplains."

"Agreed," he replied, "but you will omit to thank some, and how will their kin feel? An account in a diocesan paper is one thing; lifted and placed into permanent form in a book, omissions become glaring and offending to the relatives and friends of those omitted. Wait until this holocaust is over. Wait until all official records, now closed by military censorship, may be handed to a competent compiler who will prepare them properly for public revelation."

He was very kind, this priestly friend.

And I smiled at him as I replied, "I do not want to be guilty of 'hurt feelings,'" and then wholly ignoring the implication in his use of the word "competent," I meekly added, "but I must say thank you or — well, or hurt my own feelings!"

Perhaps it would have been better to risk the latter, for this task has been a disturbing one. Without the assistance of the Catholic press of America this work could not *be* at all, but it has been difficult to choose from the hundreds and hundreds of clippings which have been sent me. This compilation has had to be made in a hurry, too, and possibly a coma has been missed here and there. But I beg that you will read it in the "thank you" spirit in which it has been assembled, and do not be hurt by my sins of omission.

From strictly official sources I have been barred by military censorship. Most of these pages are in truth merely reprints from the Catholic press. But I have not, in all cases, been able to restrain my story telling habit, and should you sense that some of these accounts are fictionalized, you need not suspect the facts which stand out in them. I have filed with my publishers a dated source for every statement of fact; to attach these here for you would increase the size of this book and be, I am sure, an annoyance as you travel around the globe with our Chaplains.

Forgive me, then, if your "whom I know personally" Chaplain does not appear in these chapters. I offer this work as a tribute to all our Catholic Chaplains: and if it truly be "dynamite" may it serve to blast us all, Chaplains and laity, into the Army of God for the duration of eternity.

>Sincerely in His Sacred Heart,
>
>DOROTHY FREMONT GRANT

"THANK YOU!"

CATHOLIC CHAPLAINS serving our armed forces, and Catholic missionaries, have supplied the subject matter of this book. For the most part, the Catholic Press of America has written it; with minor exceptions, the compiler has merely added a few connecting sentences.

As selections were made from the wealth of material generously contributed, a few essential facts had to be kept in mind: first, that the volume should be light enough in weight to be comfortably held by the reader; and second, that its physical scope should conform to WPB paper limitations. As a result, a great deal of data which should be included here is omitted. Further, in order to place this *partial* record before the public as promptly as possible, the terse style of the new-gatherer has been adopted. No time has been expended upon the construction of palatable phrases or a search for elaborative adjectives. Actually, herein is no more than a *savor* of the story which, in a larger volume, could be told now; and which, no doubt, will be told when the War is over and the majority of our Chaplains return to their monasteries, classrooms, and parishes.

Meanwhile, as the compiler, I am indebted to many persons and publications. It is difficult to determine the order in which to say "thank you." But since charity begins at home, I will start in my own Diocese with Dr. Patrick J. Scanlan, editor of *The Tablet* who has furnished enough assistance to "bring out a new edition every six months"; and Miss Loretta Heffernan of *The Tablet* staff. Next I am grateful to Mrs. Christopher D. Kevin, Chairman of War Activities of the Brooklyn Diocesan Council of Catholic Women, to Mrs. Alfred J. Hofmann of the Queens Council of Catholic Women and Mrs. Charles Monzet of the Nassau Council of Catholic Women for loaning me copies of letters received from our Chaplains the world over who have benefited from the Brooklyn Diocesan Council's extensive program of gratis Aid to Chaplains. Because these letters were not written for publication, the excerpts are unidentified in the text, but some of the chaplains assisted by this group include, Rev. Aloysius F. Torralba, S.J., Capt., serving with the 2nd Filipino Infantry, Rev. Philip F. Mulhern, O.P., Rev. Richard J. Hawko, Army Major, Rev. C. A. Patrick, 2nd Convalescent Hospital, Rev. Peter E. McNulty, Lt. USMS., and Rev. James E. McEvoy, U. S. Army. Because he is stationed at Kings Point, Long Island, before I leave my diocese, I must thank Rev. Leo W. Madden, Lt. Cmdr., USMS., senior chaplain of the Merchant Marine, for his account of the chaplains' work in this field.

Now venturing over to Manhattan Island, I am deeply indebted to the Rt. Rev. Msgr. Thomas J. McDonnell of the National Office of The Society for the Propagation of the Faith for extensive and valuable material on our Catholic missions and the reactions of chaplains and laity who have witnessed the result of missionary labors. The first two chapters of the "Mission Tridium" are built upon factual data furnished by this office; the third is composed from a variety of sources.

It is a matter of selection whether to remain in New York State long enough to stop at Rochester to thank the Rev. Patrick J. Flynn, editor of the *Catholic Courier* for an abundance of material, and in particular for his inspiring chapter "Brothers in Arms"; and the Rev. John J. A. O'Connor, editor of *The Evangelist*, Albany, for a quantity of data before going to Paterson, New Jersey. Here, Rev. John Forest, O.F.M., Director of St. Anthony's Guild, merits an expression of gratitude, also.

Edward J. Farren, S.J., of Woodstock College, after diligent searching acquired some of the interesting data for the introductory chapter "Why Chaplains?" From Tucson, Arizona, Rev. Francis X. Donnellan, editor of the *Arizona Catholic Register*, reports a manpower shortage of priests in his diocese, resulting in but two chaplains (the Revs. Bernard Healy and Albert Knier) up to mid-April of this year. The latter is in Northern Ireland where he reports the weather "a bit wetter" than in the desert country of Arizona. From St. Paul, the Rev. Louis A. Gales of *Catholic Digest, Timeless Topix,* and numerous other Catholic Action projects responded to my call for help with a large packet of pertinent matter. So, too, did the Rev. L. W. Seemann, editor of *The Register — La Crosse Edition*, and the Rev. B. L. Barnes, business manager of *The Catholic Messenger* published in Davenport, Iowa. The Rev. Hugh A. Donohoe supplied carefully selected and neatly prepared clippings from *The Monitor*, San Francisco, of which he is editor. And I also thank Frank M. Folsom of Philadelphia for his account of an official tour of Pacific bases which forms part of the final chapter.

Aside from the Diocesan papers already mentioned, the following have also generously contributed interest or material or both: *The Catholic Mirror*, Springfield, Mass.; *The Pilot*, Boston, Mass.; *The New World*, Chicago, Ill.; *The True Voice*, Omaha, Neb.; *The Messenger*, Des Moines, Iowa; *The Tidings*, Los Angeles, Calif.; *The Catholic Chronicle*, Toledo, Ohio; *The Duluth Register*, Duluth, Minn.; *The Texas Panhandle Register*, Amarillo, Texas; *The Southern Nebraska Register*, Lincoln, Neb.; *The Guardian*, Little Rock, Ark.; *The Witness*, Dubuque, Iowa; *The Southwest Courier*, Oklahoma City, Okla.; *The*

Register, Peoria, Ill.; *The Catholic Light*, Scranton, Pa.; *The Register*, Nashville, Tenn.; *Boys Town Times*, Boys Town, Neb.; *Our Sunday Visitor*, Huntington, Ind.; *The Indiana Catholic and Record*, Indianapolis, Ind.; *The Michigan Catholic*, Detroit, Mich.; *The Catholic Transcript*, Hartford, Conn.; *The Register*, Denver, Colo.; *The Catholic Telegraph-Register*, Cincinnati, Ohio; *The Southern Cross*, San Diego, Calif.; *The Providence Visitor*, Providence, R. I.; *The Catholic Sentinel*, Portland, Oregon; *The Western Catholic*, Springfield, Ill.; *The Holy Name Journal*, New York, N. Y.; *The Messenger*, East St. Louis, Ill.; *The Catholic News*, New York, N. Y.; *The Inland Register*, Spokane, Wash.; *The West Virginia Catholic Register*, Wheeling, W. Va., and *The St. Louis Register*, St. Louis, Mo.

To these and to all others who helped in any way I offer my grateful "thank-yous" for making this compilation possible.

D. F. G.

CONTENTS

V-Mail ... July, 1944	vii
Thank You	xi
Prayer for Peace	xvi
Brothers in Arms	1
Why Chaplains?	9
Ave Maria	21
Ora Pro Nobis	31
"Push Me Back — Save Yourselves"	33
(Aloysius Herman Schmitt)	
Wanted: A Pair of Overalls to Fit	35
(Clement M. Falter, C.PP.S.)	
"Advance Against Munda!"	38
(Neil J. Doyle)	
A Divine Tip-Off	41
(James P. Flynn)	
Convoys to God	43
Territory of the United States	45
Through the Pacific Area	55
Beyond the Atlantic	63
Mission Tridium	71
"Task Force" — Reconnaisance	73
"Task Force" — Chaplains	82
"Task Force" — Catholic Laity Under Arms	93
War Is My Parish	105
"Upon a Midnight Clear" (?)	107
Daily Routine on Guadalcanal	115
"They Sail Through the Air"	122
Passed by Censor	129
Perhaps You Know	151
Pertinent Miscellany	165
Clerical Index	181

PRAYER FOR PEACE

O Lord Jesus Christ, Who in Thy mercy heareth the prayers of sinners, pour forth, we beseech Thee, all grace and blessing upon our country and its citizens. We pray in particular for the President, for our Congress, for all our soldiers, for all who defend us in ships, whether on the seas or in the skies, for all who are suffering the hardships of war. We pray for all who are in peril or in danger. Bring us all, after the troubles of this life, into the haven of peace and re-unite us all together forever, O dear Lord, in Thy glorious, heavenly kingdom.

THE "CATHOLIC HOUR"

BROTHERS IN ARMS

By Rev. Patrick J. Flynn

When the present war is concluded, another incomparable chapter of priestly heroism will have been added to the already glorious story of the priesthood. The Apostolate of the Armed Forces, as it is served by our Catholic American Chaplains, is still very much a matter of cold statistics and official record. But back of these ever-mounting statistics and records is a soul-stirring saga of sacrifice and faith.

Priests who yesterday were teaching little children their catechism are now crawling through battle mud and sailing smoke-covered seas to save souls of fighting men on the very rim of eternity.

At the most, this book can only promise to outline as intimately as possible the heroic labors of our soldier priests in these days of total war. For the priest who has gone to war the scene has changed. His sanctuary is now a shell-ridden foxhole or the deck of a warship. The shadow of a Flying Fortress may be all that he can call a chapel. His rectory office is now a fieldtent somewhere in France or the Pacific. Christ, indeed, has gone to war.

Although this book is about Catholic Chaplains of World War II, it has no intention of glamorizing war, any more than the story of the cross intends to glorify sin.

Christianity always regrets war. Her theologians lay down strict conditions before admitting the justice of any armed conflict. The significance, however, of this book which acclaims our Catholic Chaplains is this: the Church works and prospers not because of war but in spite of it.

The Church opposes war, but she has never stood idle while one was in progress. Christ refused the sword of Peter but respected the soldier of Pilate. Though the Church condemns crime, her Chaplains haunt the prison cells. She is human enough to dread disease and death, but her priests are the angels of the sickroom. Always she stands for peace but her Apostles are the comrades of warriors.

There is nothing inconsistent in all this. Whenever souls are in danger, the Church must be on hand. It is precisely because the military life is fraught with moral as well as physical danger that the Church which fights for peace must go to war.

The Apostolate of War is full of surprises. Chaplains have learned that men who ignored God on the streets of peacetime America have called for Him in the smoke of battle. That is why the Church goes to war, and why Christ dons the battle dress.

Catholic Chaplains landed with American troops in Africa, Sicily, Italy, and France. Under fire they lifted up the wounded at Anzio and Cassino. They inched their way with the Marines who took Guadalcanal and they rushed onto the bloody beaches of Tarawa with the first wave of attackers. When ships were going down at Pearl Harbor and Tulagi, our Catholic Chaplains absolved the dying under fire. Hours before our men hit the beaches of Normandy on historic D Day, priests were parachuting down with their troops over enemy held France.

All this bespeaks a Church that is present, and this is the appeal of the Church to a nation at war. Catholic Chaplains have shown our fighting men that the Church of the Catacombs and the Middle Ages is also the Church of the present. Servicemen of World War II have seen the Catholic Church ride in jeeps and sail in warships. In the lives of her priests, they have seen the Church march through mud, worship God in foxholes, sweat in the jungle, parachute from the sky, and eat K-rations. They have seen the Church bleed with them on the field of battle.

A few weeks after General Clark's Fifth Army entered Rome, sixty-four American Catholic Chaplains were received in audience by our Holy Father. Expressing a particular pleasure at seeing the American Chaplains, Pope Pius XII stated: "Just now, in this tragic hour of human history, called from the regular life of the parish or from the calm of retirement of the student or professor, you have been caught up in the maelstrom of war and thrown into the perils of battle and the temptations of a soldier's life. No ordinary shepherds of souls are needed here. Your Bishops and religious Superiors know how immediately important and how arduous is this apostolate and they have given their best for it."

The anxiety of the Holy Father for the spiritual welfare of our fighting forces is consistent with one of the commonplace practices of Catholic tradition. Wherever war has united Christian men in a brotherhood of peril, Catholic Chaplains have marched with them into battle.

The Apostolate of Soldiers is nothing new. It is one of the oldest concerns of the Church. It begins, in fact, with Christ Himself whose charming consideration for the soldiers of Imperial Rome is so vividly portrayed in the pages of the Gospel.

Certainly our Catholic Chaplains, who have listened to the problems and worries of our doughboys during this war, can read with a smile of fresh appreciation the gospel story of the Roman Officer who appealed to the Divine Master for help. *On that day Christ became the first army chaplain in the history of Christianity.* On that day Christ set the precedent of priestly action which today sees thousands of our American priests breasting the fire of battle to help our fighting men.

The Roman officer, who held the rank of a centurion, was

the commander of the Roman garrison at Capharnaum. His servant was seriously ill and about to die. The officer went to Christ and confided in Him, "I will come and heal him," our Lord readily assured the centurion. But the officer quickly protested. It was not necessary for Christ to go to his home. No, the officer did not expect that. "Lord," he explained, "I am not worthy that Thou shouldst enter under my roof, but only say the word and my servant shall be healed."

These words of faith spoken by the Soldier of the Gospel are the words which have inspired that celebrated Communion prayer of the Holy Sacrifice of the Mass: "Lord, I am not worthy that Thou shouldst enter under my roof; but only say the word and my soul shall be healed." These words whispered countless times each day around the altars of the world were first spoken by a Roman soldier. And this soldier, to the credit of his profession, forever stands unique among the characters of the Gospel because to him alone did Christ pay this compliment — "I have not found so great faith, not even in Israel."

Even as Christ hung naked and dying it seemed to be His desire that the soldiers on guard beneath the cross should be the heirs of all that was left — His sacred garments. And as the dying Christ watched those poor pagan soldiers struggle to win His garments with the toss of dice, He saw also other days when other soldiers would struggle to win a more imperishable garment — the garment of divine grace.

When the sky darkened and nature stirred in protest at the hour of Christ's death, again by the plan of Providence a soldier was deputized to speak the Redeemer's final epitaph — "Indeed this Man was the Son of God!" And century after century wherever men have heard the story of Christ, they have bowed their heads and repeated this simple statement of faith first murmured by an awe-stricken soldier standing in the shadow of Calvary.

The climax of every soldier's life comes not when he wins a battle, but when he discovers the love of God. And so the story of the cross is climaxed when a soldier makes a discovery which has rocked the world.

There in the darkness of Calvary — the world's worst black-out — a soldier poised his spear and plunged it into the side of Christ. At that moment began the world's devotion to the Sacred Heart of Jesus. In that hour men discovered the love of God because they realized that He also had a Heart. In that hour men discovered that God could really appreciate a heart that was broken. But men learned all these things only because a soldier had pointed them all out with a spear!

Not only the pages of the Gospel but also the history of Christianity is crowded with soldierly figures whose dramatic careers clearly testify that God often walks in the wake of battles and marching armies. In the battle smoke of Pampeluna He found a Spanish officer whose leg had been smashed by a cannonball. This battle-broken Spaniard — Ignatius Loyola — became a Captain of Christ and organized the Jesuits, a battalion that has won many spiritual victories for the Church.

A little later God came upon a miserable and unhappy little soldier in the uniform of Napoleon. Jean Marie Vianney stopped in a church to pray and lost his regiment. He never did find his regiment. Instead, he became a parish priest in the mean little town of Ars. And soon the soldier whom Christ had stolen from Napoleon brought all France to His Church.

Joseph Dutton was a distinguished and brave officer in the Union Army during the Civil War. He entered a romantic but ill-fated marriage. He lived hard and fast, but at the end of a long road of dissipation stood Christ with the gift of faith — faith which led to the far off Island of Molokai. Here this once adventurous and soldierly roustabout spent forty-four years working as a Catholic layman among Father Damien's unfortunate lepers.

These are only examples, but enough to show that our Lord has a particular solicitude for men who bear arms. Our bloody quarrels, which historians politely designate as our wars, can not forestall the indefatigable search of Christ for souls and even saints.

Up to the present time over two thousand Catholic priests have exchanged their clerical garb for the uniform and battledress of our American fighting men. This is according to the best traditions of the Catholic priesthood as well as the Gospel.

Through all the ages of Christendom, priests have carried on faithfully and zealously the apostolate which Christ began on the streets of Capharnaum when He worked a miracle for a Roman soldier. They marched with the Crusaders. They rode the galleys of Don John into the smoke of Lepanto, and went with the sailors of Columbus to find a New World. When the Stars and Stripes fell on Bataan in April 1942, at least twenty Catholic Chaplains joined our brave expendables in "The March of Death."

The current role of war Chaplain is not only in keeping with the spiritual sacrifice but also with the superb patriotism which has always characterized the Catholic priesthood in America.

At "the crossroads of the world," as Times Square in New York City is popularly called, there towers above all the hustle and bustle of the Great White Way a statue that arrests the eye of every traveler and passer-by. This statue, as one might expect, does not bear the likeness of one of America's great generals, or presidents, or inventors. This striking figure which dominates the bubbling and seething mass of humanity surging through Times Square is the statue of a priest — Father Francis Duffy, Chaplain of the "Fighting 69th," famed American division of World War I.

As a soldier priest and patriot, Father Duffy has long since become legendary among our national heroes. Although he was not the only priest of World War I, nevertheless, Father Duffy is the symbol of all of them. Repeatedly during this present conflict, our fighting men have written back that their chaplain is "another Father Duffy." This is a distinctive compliment; and it means that the "Priest of Times Square" is no longer just Father Duffy, but is now the enduring symbol of every Catholic priest who has served or ever will serve the armed forces of our nation.

Father Duffy's statue standing in the very hub of our national life is America's tribute to every priest who serves God and nation in the hour of battle. In welcoming this book which attempts to honor the Catholic Chaplains of World War II, we can think of no worthier accolade than to hail them as "New Father Duffy's." This book, in fact, might have been titled, *Father Duffy Carries On,* and any American who has ever walked in Times Square would understand.

WHY CHAPLAINS?

ON AUGUST 12, 1776, the Continental Congress ratified an appointment which had been made the previous 26th of January in favor of a French-Canadian Franciscan, Père François Louis Chartier Lotbiniere. He was to minister to a regiment of his countrymen which was commanded by Colonel James Livingston then serving under General Benedict Arnold. At the time of his selection Père Lotbiniere was sixty years old. By accepting the appointment he defied the instructions of the Bishop of Quebec who had forbidden French-Canadians to assist the American cause. These instructions originated in the bishop's Catholic conscience and were supported by the articles of the Quebec act of 1774 which guaranteed broad freedoms to the Catholics of French Canada. The Continental Congress had inscribed a like guarantee, but Catholics in the Colonies remained a proscribed group. Another chaplain serving the French-Canadian Army which, despite the instructions of the Bishop of Quebec, supported the American cause, was the Jesuit, Pierre Rene Floquet. Not until June 1777 did the Continental Congress authorize the appointment of nine Protestant Chaplains who did not join the troops but remained in their parsonages to be of service when any of the twenty-seven regiments to which they were assigned passed through their localities.

That the appointment of Chaplains enjoyed the full endorsement of the Commander-in-Chief, General George Washington, is evidenced by his personal request that Father Robert Molyeux serve as chaplain to the French and Indian troops of the Continental Army stationed in and near Philadelphia.

But the first Catholic Chaplains officially commissioned by the United States Government were appointed during the Mexican War, in the spring of 1846, when the forces of General Zachary Taylor were encamped on our southern border. In May of that year a Provincial Council was held at Baltimore. As a result of some deliberations among the assembled bishops, on the 20th of the month, Bishop Hughes of New York, Bishop Portier of Mobile, and Bishop Kenrick of St. Louis, called on President Polk at the White House. During the course of this visit the President indicated his desire to commission two Catholic priests as chaplains to serve General Taylor's troops. After consultation with Father Verhaegen, S.J., of Georgetown University, the bishops presented to President Polk the names of Father John McElroy, S.J., pastor of Trinity Church in Washington, and Father Anthony Rey, S.J., of the Georgetown faculty.

Mr. W. L. Marcy, Secretary of War, brought up the matter of salary.

"You understand, gentlemen, that we have no law authorizing the appointment of chaplains," he informed them. "Your commissions will be issued by virtue of Mr. President's discretionary power," he went on to explain. "But in the past Protestant chaplains have received stipends ranging from $1,000 to $1,200 annually."

"The amount is a matter of indifference," replied Father McElroy.

"Then shall we say $1,200 each, per annum?" replied Mr. Secretary, smiling.

Naturally the question arises, since the government obviously had supplied Protestant chaplains for General Taylor's troops, why, in May, 1846, were Catholic Chaplains desired.

President Polk was very frank on this point.

"The Mexicans have an erroneous opinion," he told the visiting prelates, "that the United States is making war against their religion."

That the appointment of Fathers McElroy and Rey had more than casual significance is shown in Mr. Marcy's letter to General Taylor:

WAR DEPARTMENT
Washington, May 29, 1846.

Major General Z. Taylor,
Commanding Army of Occupation
on the Rio Grande, Texas.

Sir,

The President has been informed that much pains have been taken to alarm the religious prejudices of the Mexicans against the United States. He deems it important that their misapprehension in this respect should be corrected as far as it can be done, and for that purpose has invited the Reverend gentlemen who will hand you this communication, Mr. McElroy and Mr. Rey, of the Roman Catholic Church, to attend to the army under your command and to officiate as chaplains. Although the President cannot appoint them as chaplains, yet it is his wish that they be received in that character by you and your officers, and be respected as such and be treated with kindness and courtesy — that they should be permitted regularly to attend the soldiers of the Catholic Faith — to administer to them religious instruction, to perform divine service for such as may wish to attend whenever it can be done without interfering with their military duties, and to have free access to the sick or wounded in hospitals or elsewhere.

It is confidently believed that these gentlemen in their clerical capacity will be useful in removing the false impressions of the Mexicans in relation to the United States, and in inducing them to confide in the assurance you have already given that their religious institutions will be respected — the property of the Church protected, their worship undisturbed — and in fine all their religious rights will be in the amplest manner preserved to them. In fulfilling these objects you are desired to give these gentlemen such facilities as you may be enabled to afford, and at such times as in your judgment may be most prudent.

Very respectfully, your obedient servant,
W. L. MARCY, Secretary of War.

Father McElroy spent ten months in the military hospitals at the front, and has left this tribute to the Officers of the United States Army:

It is due to the officers of the army to say that they treated us on all occasions in the most courteous and respectful manner . . . I have never met with a more gentlemanly body of men in my life, than are the officers of our Army. The more I cultivated their acquaintance, the more I appreciated their character. Polite, affable, and free from ostentation, they are an honor to their profession and deserve well of their country.

Father Rey was the first Catholic priest to fall "in line of duty" as an official chaplain of the United States Army. Acclaimed for his gallantry under fire at the siege of Monterey, Mexican bandits murdered him as he returned south to army headquarters.

During the War Between the States Chaplains of the Union Army were commissioned as Captains of Cavalry enjoying all the benefits of that rank and to the same extent as laymen. But unfortunate dissension arose over this practice, with the result that the rank and its privileges were withdrawn and chaplains attached to the army received a monthly salary. One Catholic chaplain at this time saw fifty-two engagements, and marched with his men over winding routes from New York to Key West and from the Atlantic coast to Texas. He reported that during these marches many officers and men were received into the Church; and of all the Protestants who died of sickness or wounds, and to whom in the absence of their own chaplains he offered spiritual consolation, only two refused to embrace the Catholic Faith.

There was a high proportion of death in this civil struggle. Scientific discoveries have grown apace since 1860, and Medical and Hospital Corps have been relatively increased. Today, a casualty abroad may be entered in a U. S. Military hospital within the boundaries of continental United States in from forty-eight to seventy-two hours of his fall on the field of battle. But in 1860, more often than not, as he fell, so he lay until death released him days later. Father Corby, a chaplain of 1860, related deaths by the thousands: horse drawn ambulances were scarce and slow, gangrene did not wait.

It is a strange paradox (probably only human nature) that man's constructive mind has progressed along scientific and mechanically inventive lines and so, today, has lessened the casualties in human life and limb to a minimum which is even less than in 1917–18 in proportion to the numbers involved; yet this same mind of man, who calls himself civilized, has not progressed beyond the savage barbarism that is twentieth century war, itself.

In supplying Chaplains to her Army and Navy, the United States follows the example of all Christian countries save France, eldest "Daughter of the Church," who, long since, to make the separation of Church and State visibly factual, abolished the office and sent preachers of the Gospel into battle with a gun.

In World War II the service of the U. S. Army chaplain is primarily devoted to the practice of his vocation: the same cannot be stated with as much emphasis of the Naval chaplains. But both services may be indebted to our President, a man of sincere faith who is frankly affiliated with a Christian denomination. Because Mr. Roosevelt has recourse to his faith he has no doubt readily understood that spiritual aid is an essential to the wellbeing of man, and therefore has insisted that as much of such aid as possible be made available to the personnel of our armed forces. Today there is one chaplain assigned to approximately every twelve hundred men, whereas in 1917–18 the ratio was one to every eighteen or nineteen hundred men. The Japanese bombs which fell on Pearl Harbor planted chapels in camps from coast to coast throughout continental United States. Up to that time, these had been merely white lines on blue prints reposing in official files awaiting "M" Day, and substantial appropriations.

In times past our Chaplains, Army and Navy, have been "handy men," social secretaries, librarians, liaison personnel, public relations buffers, but now, especially in the Army, their first and foremost duty is recognized to lie with the spiritual needs of their men; they are to be exponents "of the religious motive as an incentive to right thinking and acting."

Chaplains may not be under twenty-four or over fifty years of age; they must be citizens of the United States or of a co-belligerent or friendly nation. They must be approved by their Church authority, have the necessary educational background both secular and theological, and must have had a reasonable amount of parochial experience.

At the Army Chaplains' School, Harvard University, a twenty-eight day course prepares the Chaplain for duty with the troops. During this time he must participate in drilling and other physical preparation, as well as master the essentials of the practical duties expected of him. These include leadership and administration, discipline, courtesies and customs of the service, military law, military hygiene and first aid; topography, graves' registration, military correspondence and surveys; money and property; investigation; interior guard duty, field service regulations, equipment; organizations of the Army, recreation, education and music; administrative, co-operative and supervisory duties of division, corps and Army chaplains; staff regulations, Army morale and defense against chemicals.

The indoctrination course for the Navy Chaplain, given at the College of William and Mary, covers a period of two months, two weeks of which is spent at some base for practical observation. The Navy Chaplain's duties are primarily religious, but to them are added miscellaneous responsibilities. He undertakes the supervision of ship and station libraries, assists with educational and athletic activities, recreation parties, and motion pictures. He supervises sight-seeing parties, entertainments, ship dances, and Christmas parties. He also edits and contributes to the ship or station paper; and cooperates with social welfare organizations ashore in the interest of Navy relief work.

What of the men whom our Chaplains serve? They deserve, and have, a chapter to themselves. It is enough to record here that fifty per cent of the American Army goes to Church and that Catholics hold the lead for regular attendance. According to the *United States News* the American Army is the best paid, best dressed, and best fed in all the world. Two of every three are high school or college men; two of every five are radio devotees; half the Army reads books or magazines.

But this brief survey does not take account of the potential courage of this great Army nor the concern felt by our Government for its safety. The air forces are briefed before every mission; so, too, is the infantry. Before the latter go into combat the men are told what they are attacking and why, and — so far as may be surmised — what opposition they may expect. Thus our men go about their grim task of killing other men in a methodical manner. The "inspiration" of hate has no part in their indoctrination. Entering battle as calmly as may be under dangerous circumstances, the American soldier, because he is not taught to hate, gets his job done as efficiently as possible. He is not mentally retarded as are those who enter the same battle teeming, and sometimes literally frothing, with a blind hatred for the enemy.

It is a tragic fact of our age that our present enemies, and alas, some of our allies as well, have little regard for the value of human life — and man's soul. These have substituted paganism for the worship of God. One ally has indulged in superficial surface maneuvers to placate the public of united Christian nations, while another has forbidden any man in uniform to enter a church. These men, reared to worship the state — the Japanese, the Nazi, the Communist — fight with the mad fury of fanatic zealots. Motivated by irreligion and hate they squander that which the most modern and scientific factories, equipped with the most precise of precision tools, cannot produce: human life. By contrast our Government aims to spend this irreplacable commodity prudently, and as prudently gives every man in the service, should his number turn up, an opportunity to meet the Author of human life well prepared.

Chaplain Terrence P. Finnegan relates that during a desperate battle on Guadalcanal a soldier threw himself on a Japanese grenade which landed among three members of an American mortar crew. The soldier survived the explosion but, naturally, he was horribly wounded. When he was convalescing a Chaplain asked him:

"Why did you take that hundred to one chance?"

The soldier smiled, a twinkle in his eye. "It was like this, Padre," he replied, "I was ready to die. I had gone to confession. But I didn't know whether the other fellows were ready or not."

The angelic chorus announced Christ's coming among us with the obvious truth: — "Peace on earth to men of good will" — "Peace I leave with you, My peace I give unto you" said Christ when He returned to His Father in heaven. It is this peace that the Chaplain fights for with the weapon of Faith. This is the only lasting peace.

"I'll take anything they throw at me," writes one from the jungles, "but I will be glad to get back where we belong, in a world of peace where one does not hear or see instruments of death and disaster all around him. God speed the day."

And from tropical lands a priest stationed with our garrisons vividly expresses the Chaplains' mission in these lines: "I am kept on the hop all the time trying to cover five different posts which means 160 miles of jeeping over jouncing roads and 500 miles 'up-stairs' every week as a regular salesman trying to interest and convince the men that they should carry plenty of fire insurance (against Hell)."

Another, outside the continental United States, smiles as he reports: "Already some of the officers are referring to the higher morale and ascribing various causes for it — every cause but God's grace and every name for morale but morals — just a wee change of spelling. I get a kick out of it and many a quiet laugh. I happen to know some of the real reasons behind it all and they are found on God's altar. Please pray this will continue. . . . The only One who knows how much good we are really doing is God."

This Chaplain is quite right: the only One Who knows "all the answers" is God. He knows when, where, how and why this great chastisement of His children will end. He knows just how well His children, including the leaders of nations, will have learned *the lesson* of this horrible war. But meanwhile, as He permits the purging of His children to continue, He showers many blessings upon large numbers of them, giving thousands of individual souls, now helpless in the turmoil and confusion of bedlam and blood, danger and death, the grace to place their full dependence upon Him. In His infinite mercy God is reminding His children of the only purpose of their being.

Tenderly His hand protects those who love Him. "God has certainly been with me," writes Army Chaplain, William Pixley, C.S.Sp., from the Naples area. ". . . I had a tent torn to pieces by a bursting shell twenty feet away while I was just a few feet from the tent on my way in. I have been covered with dust by a shell hitting the highway no more than ten feet from my jeep, and all that happened was a flat tire, and a sore ear drum. All in all, I say, '*Deo gratias!*' "

And Chaplain Murphy, another of the Holy Ghost Fathers, writing fom England, observes, ". . . As the tempo of attack increases, our opportunities are multiplied. More and more combat men come to the Sacraments. Some of them do not get back, but it is always occasion for a prayer of gratitude to look over the list of casualties and say, 'So-and-so was in to receive Viaticum and last Blessing before taking off this morning.' . . . It is edifying to see the number of Protestant and Jewish boys who come and ask for a blessing when I drive out to the planes before a take-off. More each day I realize that we have a chance for the biggest missionary coup in the world today."

Other than the Hand of God, what stayed Chaplain Alfred J. Guenette (Lieutenant, Army Chaplain Corps), formerly of the faculty of Assumption College, Worcester, Massachusetts, when he jumped from a plane but did not descend to earth. Father Guenette was training with the Airborne Command at Camp Mackall, near Pinehurst, North Carolina, when, with a detail of troops he attempted a jump from a training plane over the Sandhills area. His parachute pack caught on the door of the plane, and he was held fast about a foot and a half below the door. Men inside the plane did not know of his plight and paratroopers jumping from the plane could not see him. But the pilots of other ships who were following the training plane sized up the situation, radioed to the pilot of the plane and Father Guenette was pulled inside. He was uninjured and the next night he went up and completed his tenth jump.

The Most Reverend James M. Liston, Bishop of Auckland, New Zealand, writing to the Right Reverend Monsignor Michael J. Ready, General Secretary of the National Catholic Welfare Conference, states: "I feel it a duty, even as it is a pleasure, to send you some word both of the chaplains and the men whom it is our privilege to welcome to our country. The Chaplains one and all . . . have edified us greatly by their fidelity to the highest priestly standards and their complete devotedness to their work.

"There are, of course, degrees in the Catholic life of men,

but whilst some, mostly through lack of instruction, are irresponsive, great numbers amongst the officers, especially of the medical staff, and the men are a shining example to clergy and laity of all that is good; we are impressed by their spirit of prayer and reverence, and their sacramental life. Some are zealous apostles of the Faith; these for the most part, though not wholly, are officers and Catholic college and university men. What a power for good they are and what fine things they will do for their country on their return home. . . ."

The Apostles themselves, who traveled by foot, mule, or ship to plant Christianity, must rejoice that modern transportation has made it possible for their twentieth century followers to cover vast mileage, as has Father Richard J. Connelly for example. Starting from the diocese of Columbus, Ohio, as an Army Chaplain, Father Connelly, serving with the air forces in North Africa, traveled seventeen hundred miles serving the personnel in the widely separated bases. And the Apostles must also rejoice as new churches are erected in the lands adjacent to the scenes of their own labors, such as the one described by Gault McGowan for the *New York Sun*. This was built by Father (Major) James A. Carey, better known as "Father Jim," of Seton Hall College, South Orange, New Jersey, now Chief Chaplain of the United States Forces in the Middle East. It was Father Jim who thanked God for the turkey and pumpkin pie when President Roosevelt and Prime Minister Churchill sat down to an American Thanksgiving Day dinner in Cairo in 1943. The distinguished company received all the details about how Father built his "Little White Church in the Desert," which Gault McGowan described as ". . . a typical New England colonial church with wrought iron railings about the entrance and a little belfry over the doorway, and a bell that might have just finished ringing for Paul Revere."

As the stories of the succeeding pages unfold, the reader will, perhaps, understand how it is that the records of the Military Ordinariate disclose ". . . a Communion record of more than half a million a month, with some months passing the million mark . . . and the records of the week-day Mass . . . an attendance of more than a quarter of a million each month. . . . Not all servicemen are saints, but many of them shame their brothers in civilian life. Good-natured but relentless pursuit by cheerful chaplains has brought thousands back to the fold from which they had strayed in civilian life. . . . No good Chaplain is ever satisfied with results since there is always much more to be done, but every wise Catholic Chaplain sees hope for the future of America in the sound ideals and good life of the great majority of Catholic service men."

Obviously, in His mercy, God has not abandoned His children who now, innocent and guilty alike, must suffer a just chastisement of their own creation. So many of us the world over, for one reason or another, have sometime abandoned "Our Father."

AVE MARIA

PERHAPS IT is because I am personally indebted to Our Lady for many favors, including the conversion of a soul dear to me, that I am inclined to pause here to honor her. Mother of us all, patroness of our country, surely she has a place in this account of our Chaplains who depend upon her intercession. But, diligent as my search has been through hundreds of clippings pertinent to our Chaplains, references to the Blessed Mother are disappointingly few. Yet it has been foretold that through her name peace will come to the world.

It is true that there are few of our Chaplains who do not hold regular weekly novena services to Our Lady under the title of the miraculous medal, or some other of her many titles. And I am confident the record of the Brooklyn Diocesan Council of Catholic Women, which has sent thousands of medals to Chaplains on all fronts, has been matched if not surpassed by the sixty-odd such Councils in the United States, as well as by other groups of Catholic men and women and by individuals. These are all in addition to the medals dispensed by the Chaplains' Aid Society in New York. Here, incidentally, Patrick Sheehan, a seventy-seven year old "soldier" adds his "bit," too, by donating several days each week to the repair of rosaries given to the Chaplains' Aid.

The call for rosaries is a constant "repeat order" from our Chaplains everywhere. Realizing the difficulty of supplying rosaries made of the customary materials, the Holy See has permitted the use of those made wholly of string, save for the cross. Crocheted knots serve for the "beads."

Our Lady is aware that neither Chaplain nor serviceman has neglected her, and she is probably indifferent as to whether the extent of their devotions is publicized or not. While my search for such publicity is insatiable, Our Lady no doubt has a special affection for such men as Pat Morris — Francis Patrick — to whom, perhaps, she extended her hand as he approached the Great Divide.

Pat Morris organized a Rosary Club among the Seabees, Marines, and Navy men at Vella Lavella. When his group moved on to Bougainville, another sprung up at Vella Lavella. Every night members gathered to say the rosary; once a week they held a general meeting which, without urging, non-Catholics attended. Pat Morris thought this club was such a good idea that he sent a message about it to his pastor back home. He suggested if the pastor would tell his people, perhaps parents, wives or sweethearts would write their servicemen of the Rosary Club, and thus honor to Our Lady would increase on all fronts.

Stories of Our Lady's intercession — all too few — have found their way into the Catholic press. One Chaplain, attached to a South Pacific Air Squadron tells of returning to the base in an almost mortally wounded B-17. "One of the waist gunners, an Irishman, and I decided, since the wounded were resting comfortably, we would help out the pilot with a few prayers. We said the rosary together. Out of his bullet ridden cage in the tail, a Jewish gunner crawled to join us for a smoke and a chat. But when he heard what we were doing he pulled a miraculous medal from his blouse and held it tight."

When the pilot finally set down his battered plane and cut his motors the Chaplain told him how his crew had helped to bring her in. The pilot wearily smiled his appreciation as the Jewish boy added: "I always carry the

Blessed Mother with me. I know she's the person who gets us out of these scrapes."

James Patrick Sinnot Devereux, hero of Wake Island, now a prisoner of the Japanese, writing to his mother under date of January 3, 1943, mentioned his reliance on Our Lady, too. "Letters and pictures would be gratefully received by everyone, also blessed beads and prayer books for the Catholics," he told her, adding, "speaking of that, I hold rosary on Sundays when we are not working."

And apparently the Jewish lad is not the only non-Catholic boy who relies upon Our Lady to bring him through a tough spot, for a young Corporal, writing to his mother, said:

"For fourteen days we pushed though dense jungles over mountains and through swamps, routing out the Japs from their stronghold on this island. Our only trouble was a little food shortage, and on our return to the base we ambushed a Jap patrol, and finished them all off.

"But believe me, under these conditions, religion means a lot to a fellow. Never before have I appreciated the Faith as I have since leaving home, — the thicker the danger, the greater my appreciation: and I expect this to keep up when I get back. It is striking to us, too, to see the faith of the non-Catholic fellows. You may not believe it, but every last one of them with us here puts a rosary around his neck before we go out on our missions. Try to get some medals for us, Mom, if you can. The fellows are always asking for them.

"If the folks back home could see the harrowing conditions and the awful places where our chaplains have to offer Mass they would be more grateful for our nice parish churches and not kick so much when the pastor reminds them these need to be heated and lighted and that takes a lota dough. Three weeks ago our chaplain offered Mass on a log for an altar, held up by two gasoline drums, — this altar was set up alongside the graves of three hundred of ours who all died as heroes. Faith means something in a 'church' like that!"

"It takes courage to hold an outpost in a jungle night," writes another G.I. "Men crouch alone in the terrible darkness. They listen for the rustle of a leaf that should not rustle, the snap of a twig that would only snap under the careless foot of an approaching enemy. The moments never seem to pass. But our boys have the necessary courage, for in the palm of the hand, each lad tightly clutches the crucifix of his rosary. From the broken, mangled body of Christ there comes to the weary bodies of our boys the courage and strength they need."

One Army Chaplain, Father Paul Lippert, enlisted the aid of the Blessed Mother to offset the offensive pin-ups in his overseas camp. A scholastic at the Holy Ghost Mission Seminary, Norwalk, Connecticut, met the plea with an original picture of Mary Immaculate, Queen of Peace, and the seminary unit of the Catholic Students' Mission Crusade has undertaken to supply Father Lippert with sufficient copies to meet all demands.

All over the world, as the engines of American war planes roar the signal of imminent take-off on dangerous missions Catholic airmen murmur the flight prayer of *Our Lady's Knights*. Members pledge to recite at least one decade of the Rosary daily, others also pledge Holy Communion at least once a week whenever possible.

Heading this gallant organization, which includes flying men around the globe, is First Lieutenant Don McDonald of Green Bay, Wisconsin, a B-24 pilot and instructor at Blythe Field, California. Other officers are First Lieutenant Bill Duffy of Chicago, training as a bombardier on the B-29 "Superfortress," Lieutenant Cliff Long of Long Beach, intelligence officer with a fighter outfit in England, Lieutenant Dick "Red" Mainzer of Los Angeles, with a radar unit in Sardinia, and Lieutenant Jack Knellinger of Bartons Ferry, Ohio, P-38 pilot in New Guinea.

All sixteen thousand members of *Our Lady's Knights of the Skies* are aware that a Holy Hour is offered at the base chapel at Santa Ana every Thursday evening. These Knights have their own insignia, designed by Cadet Paul Bodnar of Pittsburgh, showing a kneeling cadet upon whom the Blessed Virgin is conferring knighthood.

Chaplain William J. Clasby of San Francisco, post chaplain at the Santa Ana Army Air Base fathered this organization. The following letter is one of more than fifteen thousand which he has received from *Our Lady's Knights*.

Somewhere in India

Dear Father Clasby:

I've been overseas now for about eight months, stationed in India and flying a very hazardous route over the Himalayas to China. I have lost many of my friends during this time and have had a few close shaves myself.

Sometimes I got into trouble so slowly but inevitably that I had time to say an Act of Contrition and my prayers. But there are times, like yesterday, when things happen so fast that by the time my mind stopped whirling, it would be too late. I think, therefore, that *Our Lady's Knights of the Skies* is a marvelous idea as it is a constant reminder to perform an Act of Perfect Contrition before even going into the air. My prayers always include at least one decade and we here always go to Confession and Communion as often as it is possible.

Henry C. Stevenson

Another airman dedicated his service in a special way to Our Lady.

February 2, 1942

"Dear Mother Mary,

"This is your feast day; it is also the day of my departure for the Royal Canadian Air Force. Your child, good Mother, is going to war. . . . Please guard me well. I consecrate myself to you this day. Keep me pure and always worthy of you. Help me to do my duty well and to the very end; to do it as well as possible in keeping with the holy will of your Divine Son. May we always be friends, He and I, unto death and after.

"Make me a good pilot and a brave flier. Help me to do good among my companions of the armed forces, to make you loved by them. . . . Help me to work well for the return of peace and justice on this earth, for the preparation of a new reign of charity to Jesus in this world. For that, for the salvation of my relatives and friends, of the poor Africans, I give myself entirely to you good Mother. If it should please you, that Our Lord take my life for this, I accept the sacrifice in advance . . . offer it, for me, to my friend Jesus.

"May I always mount ever higher towards you. May my plane one day mount very high, very high right up to heaven. Take me then and make me quickly enter to be near you."

Before a statue of the Blessed Virgin in the home of Pilot Officer Jean Francois Bittner of Quebec a vigil light burns day and night. His letter was deposited under the statue by the young flier before he "flew away." His earthly mother found it after the writer had been killed in the skies over North Africa.

* * *

An Army officer, acknowledging the "perfect gift" received from the Holy Name Society of his parish, wrote in appreciation:

"You know the Catholic soldier has it all over the non-Catholic lad simply because he has an 'ace in the hole' — his rosary. He usually keeps it under his pillow when living in tents; in his bedroll when under the stars; in his pack when on the march; in his stripped pack when in combat. He always has it with him: it is a tremendous morale factor. Even the "pin-up" boys wonder what it is that gives the Catholic such a lift after a short nap; a moment by himself with his pack, a sleep with one hand under his pillow, or a brief visit to the local chapel.

"What a comfort! What a boost! Take any one of the mysteries — I like the Joyful ones — and just dwell on them momentarily. Why, for centuries people have been going though this same thing in one form or another, just to foster and promulgate the truths of our religion. . . . You grab those beads and fall asleep trying to say them, completely satisfied that yours is the cause justified and therefore you must do everything in your power to bring about a proper end to the dilemma. You are a Catholic! You sleep. You are ready the next day — ready to go. You might even be happy about it.

"The latter is one thing noted particularly about our boys. When it comes to surgical operations, for instance, they seem to be ready. They don't question the ability of the surgeon, they don't whimper. They ask for the padre and let it go at that — happy in the thought that they have gone to confession and Communion. They are happy then because they have something tangible, worthwhile, satisfying — the only thing that counts.

"That's all for now — except thank's again for the Rosary, the 'perfect gift.' "

* * *

Perhaps no other man who has given his life for God and

country rests in a grave as unmarked — yet as marked — as this pilot's: —

"He had taken off the carrier on a patrol mission and his engine began spitting oil and missing badly," reported a chaplain from the Archdiocese of Los Angeles. "He made a determined effort to return immediately, but the engine lost power and dropped into the drink. We watched the plane, about five hundred yards off, gradually fill up, sink out of sight, hoping, pleading, praying that it would bob out.

"I ran aft to be a little closer, as our ship had to maintain her course and under no conditions alter speed. The bridge had already signaled an escort destroyer to the crash. I prayed, running down the deck, then spoke absolution.

"Reaching out as far as I could I gave the blessing; I repeated the form a couple of times; then I took the ritual from my blouse and began the prayers for the dying. I couldn't see: tears jammed the runway. I said the prayers from memory. His plane was his casket, the sea the cemetery, his shipmates standing helplessly about became his mourners. But he needed a grave marker. I felt in my pocket for my rosary. I had that rosary from seminary days and intended to keep it for life; I looked at it once more; kissed the crucifix for the last time and threw it overboard to make a Christian grave. A tiny spot in the Pacific holds the remains of a loved one, marked with the Sign of the Cross, fresh from God's Communion table, ready to meet his Maker. A good pilot, a grand shipmate, a true Catholic."

* * *

In the third year of World War I, Pope Benedict XV, on May 5, asked for a world crusade of prayer to Our Lady Mediatrix of Graces for the purpose of ending the destruction which threatened civilization. Eight days later the Blessed Virgin appeared to three little shepherds at Fatima in Portugal.

"To save souls," she told the children on a subsequent appearance, "the Lord desires that devotion to my Immaculate Heart be established in the world. If what I tell you is done, many souls will be saved and there will be peace. The war will end; but if they do not cease to offend the Lord, not much time will elapse, and precisely during the next Pontificate, another and more terrible one will commence. When a night illumination by an unknown light is seen, know that is the signal that God gives you that the castigation of the world for its many transgressions is at hand, through war, famine and persecutions of the Church and the Holy Father. . . .

"If my requests are heard Russia will be converted and there will be peace. Otherwise, great errors will be spread through the world, giving rise to wars and persecutions of the Church; the good will suffer martyrdom, and the Holy Father will have to suffer much; different nations will be destroyed; but, in the end, my Immaculate Heart will triumph and an era of peace will be conceded to humanity."

Our Lady's requests?

The consecration of the world to her Immaculate Heart. Pius XII has complied with this request.

Communion in reparation on the first Saturday of each month. Daily Rosary. Compliance rests with the faithful.

ORA PRO NOBIS

R.I.P.

Aloysius Herman Schmitt	Pearl Harbor
Clement M. Falter, C.PP.S.	North Africa
Neil J. Doyle	Munda
James P. Flynn	Sicily

—◆—

IT SEEMS fitting here, next to Our Lady's place in this work, to mention some of our Chaplains who now, no doubt, have beheld the Queen of Heaven. Yet in mentioning only "some," as elsewhere, omissions will be patent.

Our first casualties in the Chaplains Corps are naturally included. Father Schmitt at Pearl Harbor was the first Catholic Naval Chaplain to give his life; Father Falter, on the beach of North Africa, was the first Army Chaplain to die.

Father Neil Doyle, popular and beloved by clergy and laity, has been mourned up and down the Eastern Seaboard and well inland. The account included here of his life and death has been written by a fellow priest. Father Flynn, who died in the Sicily campaign, may not have been so well known in the eastern section of our country, but he is equally mourned in his Minnesota diocese. Here is reprinted his last letter to his parents which shows him to be a priestly priest and a devoted son. This letter, almost without exception, has stood alone in the Catholic press, unique as a self-written obituary.

To me it seems presumptuous to consider putting the following accounts "in my own words" when the words of others — some who knew the subjects of these notices well — are so much more eloquent. They appear then, as they were found in clippings from the Catholic Press. Each notice stands alone; further comment would be superfluous.

"PUSH ME BACK — SAVE YOURSELVES . . ."

"HE WAS on the *Oklahoma* when she was hit (explained Lieutenant j.g. Bill Ingram, nephew of the famous Annapolis football coach, to Navy Chaplain John P. Kelly). When the general alarm sounded he went to his station, with about a hundred and fifty other men. All hatches and ports were battened down. The ship was struck so hard and fast that she began to heel over at a terrific angle before the hatches could be opened. The Padre and the men were in the corridor, squeezed like sardines; finally an aviator managed . . . to open a hatch far enough to let a man through.

"He yelled, 'Gangway for the Padre! Let him out!' . . . young doomed sailors pushed each other aside to make way for the priest to escape. . . .

" 'Pull out the men nearest the hatch. I'll stay with the rest of my boys.' Six men escaped. The ship heeled over. . . . The Padre was with his boys."

On May 29th, 1943, Mrs. Richard Buehheit of St. Lucas, Iowa, sponsored a new vessel for the U. S. Navy, naming it *U.S.S. Schmitt* in memory of her brother, Aloysius Herman Schmitt, the "Padre" who stayed with his boys aboard the *Oklahoma* which was sent to the bottom by the Japanese on December 7, 1941, at Pearl Harbor.

Mrs. Buehheit, together with Father Schmitt's other sisters and brothers, had long since read and re-read the posthumous citation signed by the late Secretary of the Navy

Frank Knox, awarding him the U. S. Navy and Marine Corps medal for bravery:

For distinguished and sublime devotion to his fellow men aboard the USS *Oklahoma* during attack on the United States Pacific Fleet in Pearl Harbor by enemy Japanese forces on Dec. 7, 1941.

When that vessel capsized and he became entrapped along with other members of the crew, in a compartment where only a small porthole provided outlet for escape, Lieutenant Aloysius Schmitt with unselfish disregard for his own plight, assisted his shipmates through the aperture.

When they, in turn, were in process of rescuing him, and his body became wedged in the narrow opening, he, realizing that other men had come into the compartment looking for a way out, insisted that he be pushed back into the ship so that they might leave.

Calmly urging them on with a pronouncement of his blessing, he remained behind while they crawled out to safety.

With magnanimous courage and sacrifice, all in keeping with the highest traditions of the United States Naval Service, he gladly gave up his life for his country.

December 7, 1941, was the sixth anniversary of Father Schmitt's ordination which had followed his studies at the Gregorian University and the North American College in Rome. God summoned him just as he had finished the Holy Sacrifice of the Mass. More than this, when, some months later the Rev. Edward Lynch, O.M.I., Navy Chaplain, sent Father Schmitt's Breviary to the Military Ordinariate, he noted that the hero priest had left his marker at the First Vespers of the Feast of the Immaculate Conception indicating he had said Little Hours before he began preparations for what was to be his last Mass.

Wrote the Rev. William A. Maguire, well known Navy Chaplain, to Father Schmitt's sister, Sister Mary Germaine, O.S.F., ". . . The circumstances of his death are well known by eye witnesses. Before his ship sank Father Al succeeded in passing at least three of his men through an air port. In spite of the efforts of those three men, assisted by a fourth, they were unable to pull Father Al to safety. Father Al is quoted as having said: 'Push me back — I am holding up other men. Save yourselves.' He returned to the interior of the ship and continued to help other men until the end came."

"He pushed a sailor through the porthole out into the churning sea," writes an eye witness. "As the man came up he had a last glimpse of Father Schmitt's face and the gold chaplain's cross on his shoulder."

Prior to his enlistment in the Navy in June, 1939, Father Schmitt had been attached to the Diocese of Dubuque.

* * *

WANTED: A PAIR OF COVERALLS TO FIT . . .

As the mighty allied armada neared the French African coast on November 7, 1942, the Holy Sacrifice of the Mass was offered on ship after ship by numerous chaplains who accompanied the invasion forces. Among these was Chaplain Francis T. O'Leary, U.S.N., of Lowell, Massachusetts; Chaplain Patrick J. Ryan, Senior Chaplain for the African troops; Army Chaplain Clement M. Falter, C.PP.S., of Rensselaer, Indiana; and Chaplain Francis Ballinger, U.S.N., of the Archdiocese of Newark.

As the armada steadily advanced toward a pin point on the North African coast named Fadahla, the finger of God rested upon one of these Chaplains. On that bleak, tense, afternoon of November 7th, Father Clement M. Falter, as priest and victim, offered the Holy Sacrifice for the last time. Less than twenty-four hours later, Father lay on the beach — motionless.

He might possibly have saved himself if he had hung back; if he had waited until his men had secured the beachhead making it reasonably safe for him to follow. But the chance is slim: God has a schedule, an appointed time for each of us. It so happens that after Father Falter had left the transport *U.S.S. Joseph Hewes* which had brought him from the States, she was torpedoed. Aboard her he had been a cheerful passenger, at ease, confident, serene. He kept himself and the Catholic boys who so desired, ready for any eventuality the will of God might design. Naturally the men were tense, expectant and so was Father Falter: so might any Catholic be knowing he faced two grave alternatives — a battle of fire and steel, or — Eternity.

What are men's thoughts on such occasions? Do they recall home, parents, loved ones, old friends? As Father Falter advanced toward the coast of North Africa to meet the deadline God, in His wisdom and mercy, had devised, did he think, perhaps, of his classmate, Chaplain John Wilson, C.PP.S. (Captain, Army) then, and please God still, on the other side of the globe, a prisoner at Bataan?

Whatever may have been Father Falter's thoughts cannot be known. But one of the men in his charge, Sergeant Charles Marlin of Dayton, Ohio, has recorded in writing his thoughts of his Chaplain. In part, here is what he says:

"He was a real friend . . . always in good spirits. Just to have him around was invigorating and bracing. . . . Especially was this the case on the trip to North Africa. If any chaplain was a real father to the boys it was Father Falter.

"We reached the African coast just after dark. The French garrison seemingly did not know we were there until about midnight. Then the lights on shore went out suddenly and we knew we had been detected. . . . About midnight I went up to the battalion office. A number of men and officers were there and with them was Father Falter. He looked very serious . . . a conversation was going on and the boys were arguing with Father Falter that he should not go with them into the barges in the morning on their first effort to land and hold the shore. They wanted him to remain on the transport until it was safer. Father was vigorous in asserting that he would go wherever the boys went. He was not there, he said, to look after his own safety.... If the men would be in danger, he would be at their side.

"We were supposed to land on the shore from our barges at six o'clock on Sunday morning. ... We entered the landing barges ... and proceeded to the shore about three quarters of a mile away through water that was quite rough.

"Our company got into two barges; Father Falter and I were separated. The last person to whom I spoke as I left was Father, ... we had become very good friends. All the way over he was teasing me about finding a pair of coveralls for him that would fit. I was on ordnance duty, but I could not find a pair for Father. He landed simply in his Army uniform. As I stepped into my barge, he said to me: 'Goodbye Sergeant, and good luck. If you do not come back what do you want me to tell her?' He referred to my wife about whom I had often spoken to him.

"Our barges were in the water for two hours, mine was always close to Father Falter's, and we touched shore side by side. We were under fire all the while and as we neared the shore a squadron of French planes flew over and bombed us. I saw a direct hit on one barge. It just disappeared. As we touched the beach the shelling from the fortress in the town became extremely heavy.

"I had taken but a few steps in the sand when I chanced to look back in Father Falter's direction. It was very close to eight a.m. Just then a shell from a French 75mm. gun landed near Father and the group with him. I saw Father fall, and quite a few about him. ... Later I learned that Father was struck in the head and killed instantly. I could not think of going to him at the moment. The shelling grew in ferocity....

"It will be a long time before I shall forget Father Falter. He meant much to the Catholic boys ... and I was pleased and happy when I learned the Government had awarded the Purple Heart to Father posthumously. If any man ever deserved it, he did."

* * *

"ADVANCE AGAINST MUNDA!"

"He was admired by men and officers alike. (Father Neil J. Doyle of Devon, Connecticut, chaplain of the 169th Infantry.) His zeal and sacrifice won him a lasting place in their affections. When the word went forth, 'Advance against Munda!' he thrilled at the announcement. This was what he had sighed for so long — to be on the field of battle, to rush to the side of men standing on the threshold of Eternity, eager for the Passport to eternal life."

He knew occasions of terror — always the same occasions, — a Solemn High Mass. He needed room to maneuver for he had a spring-like, bounding step which from Seminary days had won for him the affectionate name of "Cowboy." This was inconvenient on the occasions mentioned. "Cowboy" Doyle approached each such with the fear he would do something wrong, even more likely, he might knock something over. But on the field of battle this same bounding stride took him swiftly to the side of dying men: — or at home in the States, to the side of anyone in distress, be that distress merely a matter of too many bundles in the arms of an old lady. No challenge of courage or charity was too great for "Cowboy."

"Up and down the Munda Trail he journeyed, urging the well, comforting the sick and wounded, anointing the dying, and burying the dead — without fear, hesitation, or anxiety for his own safety. Major Sellers, battalion commander, offered an illuminating cameo of him: 'The last time I saw him alive was in a thicket just off the Munda Trail. A soldier was kneeling beside him. His right hand was raised in absolution.'"

"... a young soldier, Francis Muli ... was with Chaplain Doyle all the way from the West Coast of the United States until that bloody day in the Munda Jungle when the brave young priest fell under a barrage of enemy fire.

"... Muli tells a story of priestly heroism — the kind of story that the adventure loving Father Doyle would have loved to hear but never dreamed would be told about himself. Upon arrival in the Pacific battle area the first job was to establish a beachhead on one of the Japanese held islands in the Russell group. Father Doyle's men were young American boys wearing army uniforms and carrying guns, but scared to death when they were faced with enemy fire for the first time.

"... those young, frightened and green recruits, who drove into the Russells, will never forget the day they made their first beachhead against the Japs. Father Doyle was under orders to stay back until his men ... had established themselves. Father was to follow only when 'his boys' had made good the beachhead. But anyone who knew the reckless bravery of Neil Doyle, also knew that 'Stay back!' was one thing that the soldier-priest could never do, especially when adventure — even perilous adventure was at hand.

"Many times as a young student he had stolen time from his more serious studies to enjoy the high moments of literature and history. I believe that he somehow regretted that he was born too late to know 'Merrie England,' sail with the Christian Fleet to Lepanto, or to ride with the immortal Crusaders. ... For Father Doyle, making a beachhead in the Russells was no little military operation. It was his dream world of books come to life. It was Lepanto. It was the Battle of Hastings. It was Bunker Hill and Gettysburg for him. That is why Neil Doyle could not 'stay back.'

"Frightened and under fire for the first time 'the boys' hit the enemy shore. Could they make good? Suddenly in front of them appeared the gaunt and gallant figure of intrepid and dauntless Chaplain Doyle! He was supposed to be in back of them. But there he was in the lead, cheering the men, picking up the wounded, leading the way. Seeing that Father Doyle was with them, Soldier Muli gripped his gun tighter and whispered, 'God bless him!'

"Then came the day that Father Doyle and his men went to the New Georgia Islands. The great Japanese air base on the southwest strip of Munda Island was their objective. . . . Things went all right and after a few days Father Doyle, Francis Muli, Salvatore Calvo and all the rest were crouching in jungle fox-holes only one mile from the Munda air base. Jap fire was raking their position. Inch by inch the Yanks were edging ahead. From fox-hole to fox-hole Father Doyle crawled with his men toward the jungle air field. To his fox-hole comrades, the soldier-priest called encouragement. His face was dirty and begrimed but always smiling. Then one of his boys lurched in his fox-hole and screamed. A Jap sniper had aimed well. Sick call in a fox-hole! Instantly and heedless of danger Father Doyle jumped from his fox-hole to go to the aid of the wounded Yank. But the Japs saw the priest and a blast of fire caught him in the leg, and Father Doyle fell in the jungle without seeing — what he must have wanted to see — his boys take the Jap air base of Munda.

"The Yanks who took Munda tried hard to save Chaplain Doyle's life . . . but in a jungle bed, close to the spot where he had fallen, the strong heart of Neil Doyle slowly came to rest. 'Tell those at home to pray for my soul . . .' were his last words. . . . Soldiers wondered why he was smiling. 'When he passed away he was still smiling. . . .' Maybe it was then that the hero-priest remembered and understood something he had read back in his student days at St. Bernard's Seminary at Rochester, New York . . .

" *'The men signed with the sign of the cross go gaily in the dark.'* "

* * *

A DIVINE TIP-OFF...

"Dearest Dad and Mother," began Rev. James P. Flynn of the diocese of Crookston, Minnesota, as he wrote a letter to his parents from an advanced base in Sicily, "Funny thing about this letter — if I am alive two weeks from now you won't receive it. In other words, if you get this letter, it will mean that Almighty God has decided I've been on earth long enough and He wants me to come up and take the examination for permanent service with Him.

"It's hard to write a letter like this — there are a million and one things I want to say, there are as many that I will remember when it is no longer possible to write you.

"It's a funny thing about this operation — but I really don't believe I will come out alive. Call it an Irishman's hunch, or a presentiment or whatever you will: . . . I believe it's Our Lord and His Blessed Mother giving me a tip to get prepared. Naturally I'll go to Confession before we sail and get absolution and whatever my friend Father Denny Moore of the 15th can give me. Furthermore in the event I do get killed you can have the scant consolation of knowing that is was 'in line of duty' not only for my country but especially for the work of Almighty God.

"If losing my life should happen to be the price I pay for the position I hold (as an Army Chaplain), then it's certainly a cheap price, for after all I'm probably better prepared to meet God as my Judge now than I have ever been, and if death must come — then far better for it to come when I'm shoulder to shoulder with these men who are fighting to preserve for our country the freedoms you and I have known and to extend these freedoms to the oppressed peoples of the world. I'd a lot rather have it come when I'm giving absolution to dying men than to have it come by a heart attack in my sleep or in an auto accident. I'd a lot rather have it happen when other men are going to know that in spite of being 'scared as hell' like the rest of them, a Catholic Priest is still going ahead and doing his work.

"You know, Dad and Mother, I'm awfully proud of the fact that I'm with these men, proud of the fact, that I, a Priest, on maneuvers under the trees; in cantonment type chapels; on the tail gates of trucks; on an 'altar' set on a couple of ammunition chests on the deck of a ship; in front of a grandstand — these and a dozen other spots — so that wherever we have gone, there the men have at least had the opportunity of going to Mass and of receiving Our Lord in Holy Communion. No matter what the price is — it has been worth it, over and over again.

"I've been a Priest to these men and in many senses of the word. God has been good to me and whether He wants me to come upstairs and check over my stewardship now or in years to come makes very little difference to me — after all, it's all in His hands. But whether life is going to be measured in days or years I'll always be grateful for the perfect opportunity I've had to help my fellow-men. Some times I haven't done too good a job — at other times everything has worked out well — but at least I've had the chance — and at a time when it was much needed.

"I could write for pages and never get said all I want to say — so, my darling mother and grand dad — 'au Revoir' as the French say, and 'adieu' — until we meet again with God. We'll never be split up — for through our prayers we'll always be together and, God willing, we'll be united at His throne some day. And why not — for that's the only reason He puts us in the world, after all.

"Good luck, God bless you — and don't cry, just pray and God will see to it that everything works out for His honor and glory and your and my eternal welfare."

CONVOYS TO GOD

IN TIME of war convoys are important. They assure, in so far as is possible, safe conduct through dangerous zones, for troops and supplies.

In the course of a natural life many danger zones are encountered. Safe conduct through these treacherous areas depends upon many factors: — free will, a state of grace, dauntless faith, priestly administrations. So in a real sense and especially in war-time, our Chaplains are convoys to God.

Their official title is sometimes written "morale officer and chaplain." Morale, however, is not enough, nor is it a "first"; hence, this title, is top heavy and lopsided. If a Chaplain is faithful to his vocation and his duty, the guidance and service he offers his men is potential morale. To most of us the word "chaplain" implies one who keeps his men aware of God and their duty toward Him. His is an office of prime importance because in all things God, Himself, is of prime importance. To make "chaplain" of secondary importance in the foregoing title, therefore, is putting *second* things *first*.

In so far as he, alone, may do so, the Chaplain is responsible for "safe conducts" — to God. Naturally his convoy work must be cooperative. The men whom the Chaplain serves must do their share. Safe conduct to God does not necessarily imply *only* a "happy death": there are many who can testify it is often hard to keep on route G O D while living. But in comparison with vague morale,

God and faith are prime intangibles. When God is served and the faith lived, morale is automatically generated. Without God and faith so-called morale is built on shifting sands and kept alive principally with injections of athletics and recreation.

Following are recounted the experiences of some of our convoys to God enlisted by the United States to safeguard the faith and morals of the armed forces here and abroad. Again space limitations are irksome, because so many more accounts should be included in this section. At best, the following can only give a sample flavor, so to say, of the work, the hardships, the risks, and the "devotion to duty" of our Chaplains.

Territory of the United States

"... OVER THE BOUNDING MAIN..."

The call went out for men and ships, and the men and ships of America answered. Mere boys flocked to the adventure of which youth never tires, and old sea dogs who had long since settled down as established citizens of their communities hastened to the seaports to offer their services to man the ships of the Merchant Marine.

And what ships they were!

Every hulk that could float or be made to float, hundreds of dilapidated, rusty old tubs resurrected from the graveyards of forgotten ships, had to be pressed into service in those first days of crisis which dawned with 1942. After hasty repairs, some paint here, a patch there, a new boiler, a second rudder, they went forth into action, wheezing and groaning and weak in the joints, but ready to serve, — ready to take supplies to men who, with courage and bare hands, were facing a modern, well equipped, and well supplied fighting force.

How many Americans realize the beating those seamen and their old ships took? Convoys were few. Adequate protection could not be given, there were no guns and no munitions for merchant ships. But whatever the risk, those bare hands had to be filled — the supplies had to go through. So the men of the Merchant Marine on slow ships, unprotected, undermanned, in danger from sea and sky went forth. Ships were sunk off Sandy Hook, outside the Golden Gate, in the Gulf of Mexico, in the waters of the St.

Lawrence. But the Merchant Marine did the job. The supplies were delivered. The losses in the first two years of the war were higher than in any branch of the armed forces.

Seamen are deeply religious. When the opportunity offers whether on shore or shipboard they are as prompt as any landlubber to pay their homage to Almighty God in a formal way as God-fearing men have done from the day when the grateful children of Israel built their altar by the Jordan. It is a human thing to do, as all men of good will know. Yet, God is everywhere and those deprived of the privilege of worshiping Him in His appointed temples are all the more aware of His existence, and of the need of His help, when they are cast alone and helpless upon a wilderness of water. Is it strange that a homesick sailor, standing his lonely watch all night under the stars, hundreds of miles from land, surrounded by vast and terrible forces, should feel God's over-powering presence and the need of His friendship? Was it only a coincidence when the Divine Master walked the earth that He chose as His most trusted friends — fishermen? He sought in His Apostles simple, child-like love, faith, and loyalty, the kind that can be depended upon even unto death. And He looked for these qualities where He knew they might be found, among the men who go down to the sea in ships, for these have always been noted for those rock-like virtues. No one knows this better than the Merchant Marine Chaplain, who, today, unfortunately is shore-bound.

Now — what was the origin of the Merchant Marine Chaplains' Corps? The War Shipping Administration wisely saw that if the moral and general welfare of the officers and seamen of the Merchant Marine were to be safeguarded, some religious guidance would be necessary, and therefore appealed to the Navy for Chaplains. But the Navy had all it could do to supply its own needs. So the Administration decided to form its own Chaplains' Corps and assign Chaplains to all training stations. Archbishop Spellman, Military Vicar of the Armed Forces, was asked to cooperate, and immediately released four priests from his own diocese, where so many had already volunteered as Chaplains for the Army and Navy.

Unlike the Chaplains of the Army and Navy, however, the Chaplains of the United States Maritime Service have to do all their work without seeing the results or participating with their men in the field of action. There is no clergyman in the Merchant Marine service who would not infinitely prefer to go to sea with his men, to share their hardships and perils, and if need be, to lay down his life with them. But the very nature of the Merchant Marine makes this impossible. The men go forth in small groups according to the size and nature of the ship: sometimes 35, sometimes 50, sometimes 60. They cannot take their Chaplains with them for there would not be enough Chaplains to go around. The work of the Chaplains therefore, becomes a routine far from glamorous, and a very intense one, too, for before the men go to sea the Chaplain must fit into an already crowded training program his religious services, his personal interviews, his instructions for the reception of the Sacraments, all in addition to many miscellaneous duties which come under the head of morale work. But, as in the armed forces, the Chaplain is primarily a spiritual father, friend, counselor, big brother and once in a while a jack of all trades.

Father Leo W. Madden (Lieutenant-Commander, ChC., USMS) tells the following anecdote which occurred at one of the training stations:

"Father," said an anxious trainee, "can you help me out? I have a liberty pass, and my girl's coming from my home town to meet me."

"Well?"

"The guard at the gate won't let me go ashore."

"Won't let you go ashore?"

"No. He says I need a haircut."

"And he is right. You do. You certainly do."

"But, Father, the barbers have left for the day."

The Chaplain reflected. He noted a barber shop was still open.

"Could you get me some clippers over there?"

In almost nothing flat the trainee was back with the clippers.

"Sit down," ordered the chaplain. Then he gave the surprised trainee a very professional haircut.

"Gee," said the lad with a liberty pass and a girl to meet him, "where did you learn to cut hair?"

"In the seminary," replied the chaplain, gravely eyeing his results. "I cut the hair of all my fellow students for four years."

It is the fond hope of the Merchant Marine Chaplains, even if haircutting must be included, that they may extend their service to the various ports of the country to meet the seaman when he comes ashore for his well earned liberty, to warn him against the pit-falls, for it is no secret that when the sailor goes ashore he is often preyed upon by unscrupulous and dishonest landlubbers: it is wholly possible that a port may be a greater danger for him than a raging tempest or battle at sea.

Shore-bound though they must be, the Chaplains of the Merchant Marine have found the spark of adventure in the eyes of those American youths who report to their training stations a contagious affair. They see the souls of these lads, the most precious cargoes ever borne — the souls for which Christ died. They see these souls endangered, separated from their homes and their loved ones to overcome worldly enemies. They hope to send all these boys out to sea as they sent Edward Gavin. . . .

— Courtesy Holy Ghost Fathers, Norwalk, Conn.

The "Pinup Madonna." Painted by a Seminarist.

Prayer before Makin Attack. Chaplain L. W. Yarwood (Lt. Col., U.S.A.) says Mass on a transport. An ammunition chest forms the base for the altar.

— Press Association, Inc.

— Courtesy the Brooklyn Tablet.

— Courtesy Office of Chief of Chaplains.

Chaplain Thomas M. Reardon (Lt., U.S.N.), the first Chaplain on Guadalcanal.

Chaplain Stephen J. Meany, S.J. (Capt., U.S.A.).

New Georgia. Chaplain Paul J. Redmond, O.P. (Lt. U.S.N.) blesses a group of Marine Raiders before their push across the island.

—Official U. S. Marine Corps Photo

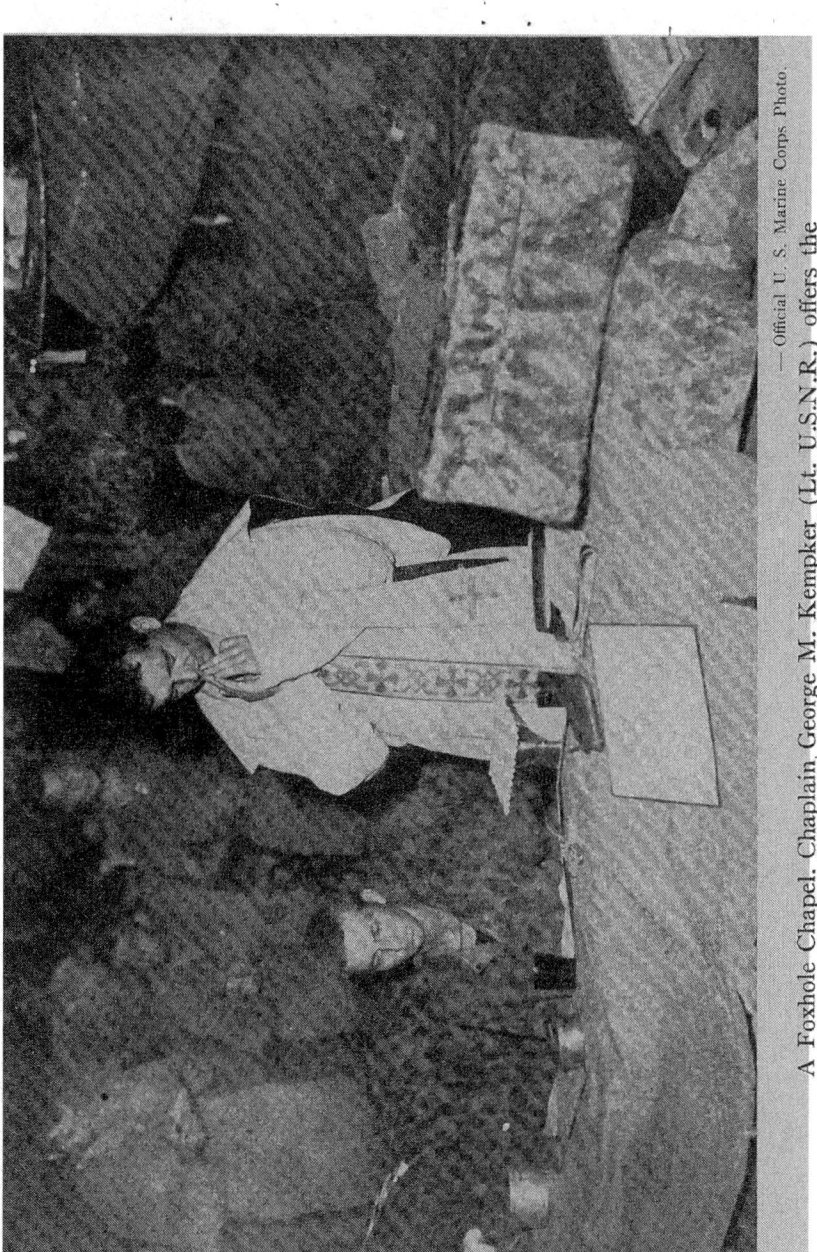

A Foxhole Chapel. Chaplain George M. Kempker (Lt. U.S.N.R.) offers the Holy Sacrifice for his Marines on Bougainville.

— Official U. S. Marine Corps Photo.

Mass in Sicily. The hood of a jeep forms the altar of Sacrifice for this unidentified Chaplain.

— Signal Corps Photo.

Life Amidst Death. A Marine detachment in the Marshalls assist at Mass for their fallen comrades.

Mass Amidst Ruins. A wrecked radio-station power house, on Saipan, becomes a chapel for Chaplain C. C. Riedel, C.S.V.

— Official U. S. Marine Corps Photo.

"Dear Father: — Your kind letter of condolence has been received. The news of Edward's loss was a great blow. But I do not worry. My Edward was a good boy. He died brave and un-afraid because he was in a state of Grace.
A Heart-broken Mother,
Mrs. Linda Gavin."

It is the prayer of the Catholic Chaplain of the Merchant Marine, a prayer not so much uttered with the lips as lived each day as he goes about his manifold duties, that if his boys are to die, they die as Edward Gavin did, in a State of Grace — and that if they are destined to return to port, that they may come, as they went forth — in a State of Grace.

* * *

REDUCING THE WAIST LINE

(Excerpts from a letter written by Father Thomas E. Foster, Chaplain at the Amarillo Army Air Field.)

"My arrival at Amarillo Army Air Field was greeted with a terrific wind storm and I was thoroughly baptized in the famous 'sands of Amarillo.' This was in February and the winds — typical of that period — were so forceful that my automobile door was nearly ripped from its hinges. Amarillo is in the Panhandle section of Texas — cow country of rolling plains on which there isn't a single tree for miles. In summer it is unbearably hot. In winter bitter cold. Texas is not a state that does things half-way! The Army Air Field is eleven miles outside the city of Amarillo and appears to have been deposited intact upon the bosom of the plains rather than to have been built there. Each building on the field has guy lines at the four corners to anchor it securely. Dust blows steadily throughout the year, and I frequently awake in the morning to find my bed, pillow and me coated a brownish-red. We are famous for our rains, too. When it rains it comes in torrents and we often see our wooden sidewalks floating down the street. A phenomenon occurs when we are having one of our dust storms and it suddenly begins to rain. Then it literally 'rains mud.'

"The Technical Training Schools give the enlisted men a thorough foundation in airplane mechanics, but the going is tough. The boys go to school in shifts, stretching through the twenty-four hours. Nothing is permitted to interfere with schedule. When a man's day off arrives, he has earned it.

"We have five chapels on the field — all strictly G.I., but containing all the essentials. It taxes the ingenuity of us all, but it is interesting and invigorating. We took over one of the chapels, made it a Catholic chapel, (reserved for the Blessed Sacrament), and erected Stations of the Cross. We appealed to Father Drury, the young Irish pastor of the Cathedral in Amarillo, and he gave us a set of vestments. We bought other necessary articles and we proudly proclaimed ourselves a Catholic Church. We offered nine Masses on Sunday, each of the Catholic Chaplains saying three Masses and the priests from Amarillo saying three. Our waist lines grew thin because our daily schedule permitted only two meals on week-days and one meal on Sundays at 5:30 p.m.

"In May the other Catholic Chaplain, Father Mroczka was called to foreign duty. Six weeks later my classmate, Father Norkett, . . . came here from the Chaplain's School at Harvard. Then the first of August Father Mulligan, a Passionist from the East, arrived and we had, for the first time, enough Catholic Chaplains to care for the spiritual welfare of the Catholic soldiers.

"In the early part of July we began our campaign for

the establishment of a Holy Name Society on the Field. We called a meeting of the Catholic commissioned and non-commissioned officers and laid our plans. We enlisted the aid of Bishop FitzSimon. On August 29th, the campaign culminated in our first Holy Name Communion Sunday. Bishop FitzSimon offered the Mass. We enrolled 1250 men that day to whom we distributed Communion. Since then we have enrolled another five hundred. Now, since the Holy Name Society has been established at the Field, the number of Confessions and Communions has increased more than fifty percent.

". . . It is our plan to have two Miraculous Medal Novena services every Tuesday, one in the morning for those who are going to school at night, and one in the evening for those who are going to school during the day. Our daily Masses are at 8:30 a.m. and 6:30 p.m. at one section of the Field, and at 7:30 p.m., at the other section. We have about thirty men at each Mass and a good number receive Holy Communion. For some, it is impossible to attend daily Mass. It is most gratifying to see men coming in to visit the Blessed Sacrament at all hours of the day and night. But there is a great deal to be done and we have concluded that it can't be done by the Catholic Chaplains alone. The people back home have no realization of how much we need their prayers for the success of our work in the Army among the soldiers. While there are thousands living up to the Faith, there are still a like number of thousands failing to practice it.

"An interesting side-light on the Brotherhood of Army Chaplains is this incident. I was to take twenty men to the Cathedral in Amarillo to receive Confirmation. Unfortunately, at the last minute, I could not go, so two Protestant Chaplains took the boys in to the Cathedral, stayed for the entire ceremony and brought them back in time for the next military formation.

"Visiting the Italian Prison Camp located near here, I met two Italian chaplains. Their English vocabulary extended to 'Hokay,' 'Goot morning,' 'Come on,' and a few other similar phrases. So we turned to the language of the Church. It brought more strongly to mind the 'Universal brotherhood of Christ.'

"The prisoners have turned their hand to making various religious articles from tin cans. These men are accomplished tinsmiths and the things they fashion from the lowly tin can are amazingly clever.

"I am assigned to a Technical School Squadron as that Squadron's 'official' chaplain, and naturally, I try my best to spend as much time with the boys, participating in their activities, as possible. One incident, however, proved quite an ordeal. A group of 'my' boys were going on a cross-country hike which was to emulate actual war conditions. We were informed that when the alarm was given we were to seek cover along the roadside in the deepest hole or heaviest brush possible. Wanting to set a good example, when the alarm sounded I dived into a large clump of sage brush. My zeal was rewarded with a bath of thorns — the bushes hid a patch of cactus; and I emerged covered with thorns. The soldiers laughed as they helped me de-thorn myself."

* * *

. . . A LOT OF MEN IN ON THIS MASS

"I have never seen or heard, much less offered a dialogue Mass," said Father George Casey an Army chaplain from Dorchester, Mass., "But one night I offered a Mass that was as much a community affair as a Mass could be: and into which the congregation entered as intimately and as importantly as any congregation in the catacombs.

"We were bivouacked in a cotton field on the outskirts of an immense, sprawling camp, to which we had come for firing on the rifle range. We had arranged for Mass at sundown, because the men had to get their firing done, their rifles cleaned, their supper eaten all before darkness fell.

"A detail of five men who had previously selected the site, volunteered to tie up a tarpaulin between two trees at the top of a gentle slope, so that the Mass would be sheltered. Two others carried the heavy Mass kit from the ammunition shed, a long way over uneven cotton rows, and got busy setting it up on an altar improvised by another group. Meanwhile, the soldiers, not in the least resembling apostles in appearance, sweaty beards, muddy fatigues, camouflaged helmets, but quite apostolic in purpose, ran out through the company streets, calling in to KP's, 'Hey, any you guys Catholics? Mass at 6:30; Mass over near the 247th mess. Father wants all the Catholics at Mass. Come on!'

"Finally priest, people and material get assembled and Mass begins. Two shaggy soldiers kneel by his side and mumble something that once was Latin. The breeze whips around the canvas, so a man on each side gets up, without suggestion, and holds on to the flapping altar cloths. The breeze flutters the pages of the Missal as well, so another, taking good example, climbs off his knees and assigns himself to holding down the pages. The wind spills in over the top, tearing at the candles, and by a happy inspiration two more of the men slip around back of the altar and cup their hands behind the flames.

"The Mass had not gotten under way very punctually: it is later than one might think, darkness has gathered and Father's nose is almost on the book trying to discern the words. The congregation stirs uneasily; the altar boys squirm and twist and grimace. . . . It works. One fellow, with a delighted grin slaps his hip; sure enough, a flashlight, and quickly he jumps to his feet, gets right behind Father, and holds the light over his shoulder. There were a lot of men in on this Mass.

"You would think that all this would add up to a lot of distraction. But I thought of those old Masses where everybody took part. Too, twilight is a quiet time, and that Mass was quiet and absorbed. The camp fell quiet and the men still cleaning their rifles did it softly; the KP's didn't bang the pots and pans.

"As the shadows deepened, a slender, silver sickle of a moon with an emerging star within its arc and which seemed to have been waiting just there for just this — came out just above the altar. And all things there, the moon and the stars, and the trees and the trucks, and the tents and the stacks of rifles seemed to bow towards the uplifted Host when the bell called them to attention.

"The Mass does its own preaching. When I drew apart to unvest, man after man came over to me seeking the Sacrament of Penance. And I sat upon a fallen tree, and heard them; and I never saw moon or stars brighter than those above me."

Through the Pacific Area

"BURNED, BLINDED, AND DYING..."

It happened during the battle of Lunga Point that a Japanese shell sent Navy Chaplain Arthur C. McQuaid (Order of the Purple Heart) ashore to die.

Father McQuaid was one of the first priests from Massachusetts to enlist as a Lieutenant in the Navy. He was put ashore at Tulagi Harbor where the doctors in utter dejection shook their heads and mournfully announced he would never go to sea again — except to cross the bar of the Great Unknown. But, due to the modern "miracle" of blood plasma Father McQuaid fooled the doctors and the Japs and may be found any day in a parish rectory in Lowell, Massachusetts.

The ship to which he was assigned had orders to intercept a Japanese convoy which was rushing reinforcements and supplies to their troops who were still making life miserable for our men on Guadalcanal. Father's ship was one of a task force which chalked up two large and four smaller destroyers, two troop transports and a cargo ship, as well as seven thousand Nipponese souls who, for the glory of their Emperor, descendant of the Sun-god, were pledged to destroy the lives of our sons, fathers, brothers, and husbands.

Naturally these casualties not only annoyed the enemy, but seriously inconvenienced him, and so he "slapped back" — and hard, too. Father McQuaid, aboard his ship, was at a battle station specifically chosen because "it was handiest to all the guns." During this fierce battle of Lunga Point a

Japanese shell hit the ship. The fire, heat, and smoke from this exploding shell felled many of the crew. They lay on the deck, burned, bleeding, and dying men, Father McQuaid among them. Some hours later he was lowered over the side, put ashore at Tulagi and bidden a last farewell.

His own words continue the story, which, praise God, has a "happy ending" . . .

"I was badly burned on the face, arms and legs. I went through two weeks of blindness, two months of being unable to talk, and am now regaining the use of my hands. I haven't a single doubt but that without blood plasma I wouldn't be here now. There was plasma aboard my ship and I remember receiving my first pint in the sick bay. In spite of what that did for me I was lowered over the side of the ship after the battle, neatly wrapped in about one hundred yards of gauze bandage, and set ashore at Tulagi to die. The medical officer must have misjudged the power of plasma. There is nothing like it. I don't remember getting more but my medical record shows that I was given two more units. I wish I knew who the three people were who donated the blood that made the plasma that saved my life. I'd like to thank them personally. I am grateful to all of them and so is every man in the Navy, Marine Corps, Coast Guard, and Army who has lived because of blood plasma."

* * *

"PADRE INC."

Somewhere in the Russell Islands a zealous Catholic priest from Woonsocket, Rhode Island, erected an eye-catching sign before his jungle headquarters:—

PADRE INC.
John P. McGuire, Senior Partner
John F. Culliton, Junior Partner

Father (Lieutenant-Colonel) McGuire was one of the

three priests who ministered to Father Neil Doyle after he was mortally wounded. Father John P. Mahoney was the first to reach Father Doyle, giving him spiritual as well as corporal first aid. Then, at the Clearing Station, Father McGuire, noticing a mist gathering over the wounded Chaplain's eyes, asked, "Would you like to receive the Sacraments, Neil?" At the base hospital Father McGuire's Junior Partner, Father Culliton, met the dying priest and remained with him until the end.

These few paragraphs relate to some extra duty the Junior Partner, who hails from the diocese of Hartford, acquired.

There was a U. S. Navy Construction Battalion recruited from the Atlantic Seaboard states ranging from Maine to Pennsylvania and predominantly Catholic, which found itself on heavy duty in the South Pacific and minus a Chaplain.

Time and again a priest, and therefore the Sacraments, were just within reach or at least within range of vision, when sailing orders removed these Seabees from their most ardent desire.

For example: having shipped from the United States without a Chaplain, they called at Panama on Christmas Eve, but the ship did not dock long enough for shore liberty. There was no Christmas Mass and Communion for the Seabees, for their ship was going through the Canal. For days and weeks she slowly plowed her way across the vast Pacific, finally dropping anchor one day at a large South Pacific Island. From the deck a large Cathedral was visible to every eye of a Seabee. And shore liberty was granted. The Cathedral was "invaded" by the U. S. Navy. Arrangements were made for shore boats in time for Sunday Mass. But, came Saturday afternoon and the ship weighed anchor, charting a northerly course.

Within the following week the construction battalion

finally reached its "military objective." The landing was made on a weekend. There was a rush, a hustle and a bustle to settle the camp, to get through appointed tasks. Why? Rumor had it a Chaplain could be procured and would hear confessions at a chapel outside the camp area. Work completed, that Saturday afternoon the Seabees hiked off to the chapel. It was there. That much of the rumor was correct. But the Chaplain would not arrive, according to a note on the door, until seven next morning, Sunday, to offer the Holy Sacrifice.

The Seabees returned to camp. "Next morning — tomorrow, *sure!*"

By seven o'clock that Sunday morning the line-up for Confession was so long that the startled missionary priest first explaining the obligation involved, gave all general absolution. Then, *Deo Gratias*, after a journey of months covering eight thousand odd miles, the Seabees received their Lord in Holy Communion. Thereafter for a month daily Mass at the little chapel was well attended.

But then their mission in that place was accomplished and the construction men moved on again, and again the Seabees were without a Chaplain and the Sacraments. Their next "location" was five miles from a chapel — yes, yes, of course, a truck could take them to Mass — but, no, sorry, the work in hand is too urgent, no transportation can be spared.

And then, at long last, when every face of a Seabee was long and drawn, Father John Culliton arrived on the scene. He had not been assigned to the Seabees. No. As Junior Partner of *Padre Inc.* he was Assistant Division Chaplain, because his Senior Partner, Father McGuire was Division Catholic Priest. But the Personnel Officer of the Seabees, in a manner of speaking, detained Chaplain Culliton in "protective custody" until he promised to apply this same security to the souls of the Seabees.

He promised.

"But I want a chapel," he said.

He had a chapel.

There were so many volunteers to build it in their free hours that it was ready for the astonished Junior Partner in two days flat.

Then life became normal again for the Seabees — as normal as life may be in time of war and far from home. Confessions were heard regularly on Saturdays, followed by a Novena service and the recitation of the Rosary. Mass was offered at six each Sunday. The Seabees overflowed the chapel; almost to a man they received Communion weekly.

"I think we should name the chapel after a construction man," said Father Culliton the first Sunday.

When he left the building after Mass, the name, hurriedly painted, was nailed over the doorway. Father Culliton shaded his eyes to read it as, still glistening wet it reflected the brilliant sun: — "St. Joseph's Chapel."

The Seabees are not the only "extra duty" for the Junior Partner of *Padre Inc.* He also takes care of the personnel of a group of P.T. Boats. His official assignment is with an Army unit.

Like his lamented, adventure-loving friend Father Neil Doyle, Father Culliton has had his taste of fox-hole parochial duties, too. War Correspondent Jack Mahan has told the story as it occurred when the Japanese on New Georgia Island suddenly attacked a command post:

"Our fox-hole was on the southern rim of the camp. We had no time to remove the jagged rocks on the floor of the hole. . . . The screams of the enemy's bullets and the answering rattle of machine-gun fire to our right and left soon made us forget our discomfort. Japs infiltrated the camp with hand grenades and attempted to learn the Americans' positions by calling in English. Cries for help went unanswered as the troops had strict orders not even to look up when calls were heard. . . . The two sergeants with us clutched their automatics. The padre and I clutched the dirt. Chaplains and correspondents are not permitted to carry arms. All we had was a prayer."

A Jap with half his stomach shot away moaned for hours only fifteen yards away from Father Culliton, and died in the night, calling for help in English. During the night an American sergeant came through Father Culliton's foxhole on the way to his machine gun post with extra ammunition. Father begged him not to go on. "I got to get back to my men" he said, and went on — to die in a cross fire of machine guns.

Suddenly the night exploded with artillery fire. The first salvos landed about thirty yards from the command post and the men in the fox-hole feared that the Japs had begun to shell the camp from the sea. But it was American artillery fire from an adjoining island which soon got range of the Japanese positions and when daylight came Father Culliton and the men were able to come out of their shelter and move around.

So Father Culliton began the duties of another day. The Junior Partner is a busy man.

* * *

"HE SEEMS TO BE RATHER SHY . . ."

The Navy gave him the Silver Star and the Order of the Purple Heart. The Navy advanced him to the rank of Lieutenant Commander. God gave him his life.

"Yesterday I had the surprise of my life," wrote Seaman Joseph C. Maid of Rochester, N. Y. to his mother on July 29, 1943. Young Maid was attached to a Naval Mobile Hospital. "Some new officers had come in. . . . When the rush of serving them was over I saw the back of a head that I recognized. He called for coffee. When I gave it to him he kept looking at me as if he knew me. . . . 'Are you a chaplain?' I asked. His answer was 'yes.' 'Are you Father Wheaton?' He said he was! The very same who was assistant pastor at Holy Rosary parish back home!

"It's funny, though: I have served him every day, but we haven't talked much since that first time. He seems rather shy . . ."

Perhaps "shy" is not the word for Father John K. Wheaton now at Sampson Naval Training Station. Four foodless days on a raft which was dropped from a plane as tired, injured men floundered and prayed in an open sea does not necessarily make one shy. Ten days ashore on enemy held Vella-Lavella Island, hidden by friendly natives until one's own countrymen in a daring moonlight sortie effect a successful rescue, does not make one shy. When God, through the implements of a rubber raft, friendly strangers and a brave rescue hands back the life which was almost lost, the survivor is not shy, but rather hushed and humbled as he reflects upon the infinite mercy of his Father in Heaven.

When a Japanese torpedo fatally pierced the hull of the *U.S.S. Helena* off Kolombangara Island in the Kula Gulf, Father Wheaton, alone at his post of duty — his battle station in the sick bay — had to swim to the stairway before he could get above deck. There he heard the order to "Abandon ship!" — and obeyed.

"I came pretty close to going!" is his pithy comment about the hours which followed in the open water.

"At such a time," he observed, "everyone prays and prays hard: even the fellow who never went to church prays, perhaps for the first time."

Shy? The open sea, the raft, the enemy held island, — all were odds in favor of death. But the Hand of God swept the odds aside. No, one then is not shy, but cloaked with a new humility. Awesomely, silently one wonders why God has saved "me" when thousands have died. To chatter about one's deliverance would be a false note, like the screech of a fingernail on a blackboard. Rather from the depths of a grateful heart comes a whispered "thank you" and the soul prays, "Please, dear God, what will you have of me as my next assignment?"

Beyond the Atlantic ...

A NAZI MINE IN ITALY

FATHER ALBERT J. HOFFMANN of the Army, and the Archdiocese of Dubuque, has been awarded the Purple Heart, the Silver Star and the Distinguished Service Cross for bravery in action in North Africa. This is the gallant chaplain, devoted to his duty, who completed a grim task which proved too much for a burial party. A body had lain so long before such a party reached it that the detail about faced without touching it. It was Father Hoffmann who picked it up, and carried it to a waiting jeep which took the Chaplain and his burden to the new cemetery.

And then one day it happened. Father stepped on a Nazi mine. When eventually he was evacuated back to the United States he came minus the foot and leg which unknowingly had taken that fatal step. His parents first heard of the accident, through the following letter:

Somewhere in Italy,
November 16, 1943

"Dear Mr. and Mrs. Hoffmann,
Dubuque, Iowa.

"I am writing this letter at the request of your son, Father Hoffmann.... I am his assistant, and have been since I was in Ireland....

"He was wounded a week ago by accidentally stepping on an enemy mine.... His one foot was so badly hurt that it had to be amputated. He also got a few minor scratches or wounds elsewhere on his body, but none serious. He is being taken care of in a nice American hospital in our rear area. I am afraid he will not be back with this regiment and I know that we will all miss him very much.

"In his recent letters, he tells me he has written you of continuous front line service from Tunisia over here to Italy where the going has been tough as we battle the German troops from mountain to mountain. One time Father was at the front for five days without once being able to take off his shoes to rest: and there was a day and a half without food, too.

"Father tells me he always thought the stories about German atrocities were 'baloney' but since the invasion of Italy he knows better. He was with the first of us who entered a small Italian town and he went right to the Church. There he found the tabernacle broken open and filled with rubbish. The chalice was broken and the candlesticks, too. The Stations of the Cross were thrown over onto the floor and the vestments were torn and scattered all over. What the Germans do not take with them, they destroy.

"But they take everything of value, even clothing and household furnishings. When the owner of a house objects he is given a piece of paper which says that Badoglio and the Americans will pay for everything. The morale of the German soldiers in Italy is much lower than those we fought in Africa. They seem to know they cannot win but they are determined to put up a good fight.

"If it is possible I shall go to visit Father again. He is not in a serious condition, but his hand is damaged and it is just now not convenient for him to write to you. I will write again and report his progress to you. I am glad to write to you for Father, but I am sorry for the news I must put in this letter.

"Sincerely yours,
"Raymond Sauer, Pfc."

While Pfc. Raymond Sauer was penning subsequent letters, in another part of the same sector, General Clark was also writing about Chaplain Hoffmann, addressing his words to Archbishop Francis J. Spellman: —

"It is with regret that I write to inform you of the departure from this theatre, due to wounds received in action, of Chaplain Albert J. Hoffman, former Regimental Chaplain of one of our infantry regiments. You will, I am sure, be proud to know that Chaplain Hoffmann exemplified in a singular degree the qualities that make a chaplain valuable to our Army. For long months before his unit was committed to action, he studied the art of the infantryman, that he might be better fitted to serve his men. In him were blended skill, courage, and understanding of the splendid possibilities of spiritual work among our troops. During the campaign in Tunisia, these qualities did not escape notice: he was awarded the Silver Star for gallantry in action at that time.

"When he was wounded, he was engaged in religious duties on the field of action. I want to inform you personally that the contribution of this exemplary priest toward victory was great. May I ask you to convey to Chaplain Hoffmann my appreciation for his devoted service? I should be pleased if you would convey my views also to his Superior in civilian life, Most Rev. Francis J. Beckman, Archbishop of Dubuque."

Back in this country Chaplain Hoffmann spent some time at Stark General Hospital, Charleston, South Carolina. He was asked to relate the event leading to the award of the Silver Star, but he tossed the question aside, remarking, "In combat no one stands out as doing anything heroic. Acts of heroism are commonplace. Probably the only reason anyone gets a medal is that his deeds happen to be noticed and reported. As to my spending most of my time in the front lines with the men — well, this is the way I look at it: the men brought to the aid stations are usually under morphine. They will be cared for by the Chaplains in the hospitals. The fellows wounded at the front, perhaps lying for hours before help reaches them, are the ones who need a Chaplain. There is nothing more terrifying than the feeling of lying alone, lost and helpless. Those are the men whom I have made my particular concern."

This was all Chaplain Hoffmann would say — but there were other patients in that hospital, glad to talk of him. One related: "Our battalion was ordered to take hill 490, the smaller hill near 609. We got halfway up when enemy fire forced us to take cover. One of the fellows up ahead got hit. We could hear him moaning. Two medics tried to reach him, but they could not because of enemy gunfire raking the area. The poor guy kept calling and the two medics took another stab at it, but they could not make it. Then Father Hoffman walked up there through a hail of machine gun bullets and in a little while he came back carrying the wounded man."

And still another patient said that Father Hoffmann had rescued him after shrapnel had made his legs useless. "It took the Padre several hours, and the two medics with him, but rescue me he did, and under fire all the time, too. Before the night was over Father brought in four other wounded, and one of them was a German prisoner."

— God bless you, Father Hoffmann!

* * *

"GARBAGE" COLLECTOR A LIEUTENANT COLONEL

When Chaplain Henry F. Ford (Lieutenant Colonel, U. S. A.) returned home on furlough the week of March 26, 1944, his home town paper, the *Denver Catholic Register* ran a banner headline: CHAPLAIN "GARBAGE" COLLECTOR FOR ORPHANS, IS HOME. The Rev. John

Cavanaugh, who interviewed Chaplain Ford for the *Register*, quickly explained, under that startling headline, that the collecting was done in Naples after Father Ford had seen the hungry orphans at the homes conducted by the Sisters of St. Joseph and the Sisters of Mary.

"The task required the commandeering of an army truck and the extraordinary good will of several mess sergeants. As a matter of fact the soldiers co-operated to such an extent that it became customary for the commissary to consider the orphans a part of the camp personnel when orders were placed for supplies. In the kitchens the cooks estimated the amount of food necessary for the soldiers and regularly were heard to say: 'It will take more than that if Chaplain Ford's orphans are to eat tonight. Better peel another bushel of potatoes.' Even more significant, and a military phenomenon, the men on KP actually peeled the spuds with a smile."

Chaplain Ford met many new experiences on his overseas assignment. A friend of the late Cardinal Hinsley, he preached in Westminster Cathedral, London; he lived through bombings and strafings with his men; he was torpedoed at sea; he witnessed the mangled bodies of soldiers restored at base hospitals to a reasonable facsimile of their former selves; he was cited by the First Regiment of Zouaves, but according to his interviewer in the *Register* his "greatest thrill" came from the feeding of the orphans. This is his most memorable experience of his foreign service, and the only one which he freely discusses. "It often required extreme ingenuity to maintain my daily 'scavenger hunt,'" he explained, "especially when I tried to assemble a 'balanced diet.' Sometimes a crate of oranges or apples would somehow get mixed up with the other food on the truck, and on one occasion I was 'surprised' to find in the orphanage a large box of wieners. Before I could rescue these the children had utterly destroyed their salvage value."

Father Ford returned to America with a high regard for the power of our enemy and the thoroughness of the Nazi machine in such details as the mining of the Anzio beachhead.

The correspondent for the *Register* was a persistent news gatherer, for he elicited from Father Ford the information that in Africa the natives indulged in robbing the graves of the soldier dead. The "loot" sought was clothing. But the English soon put a stop to this desecration by the simple expedient of mining the graves. After a considerable number of the thieves were blown to bits the practice ceased.

Chaplain Ford has now been reassigned to duty in this country where, praise God, he no longer has to hold his breath when a bomb ceases to whistle: "that is the moment when you have reason to fear a bomb."

* * *

THE CHURCH UNIVERSAL ...

(Not *beyond* the Atlantic, but *on* it.)

This is not a story of heroism, of bravery under fire, of citations, awards and decorations. It is not a story of conversions, of the restoration of fallen-aways, nor a story of gallant death under fire. It is only the story of a "special day" and a Mass of Thanksgiving. A day which held its secret but which was not lonely because the Church is universal. Even the name of the central figure, a Chaplain from the diocese of Los Angeles, is not known to this recorder. I shall call him Father Jim.

Father Jim was coming home. He had been in the "European theater" for two years. Today he was coming home, but for yet another reason it was a very "special day." Alack, there was no one aboard with whom he felt inclined to share his secret. There were former shipmates, an assortment of persons he had met in the course of his wanderings over the seven seas. There were soldiers he had comforted in base hospitals, sailors he had met at mobile hospital units — but no one — certainly not among the Polish refugees who were going to the States — no one with whom to share this "special day."

But he could offer Mass, a Mass of Thanksgiving. Though they would be several days at sea Father Jim, for this very special occasion, took out his last set of clean linens. He rigged his altar as usual on the weather deck: the sea was very rough. It was after he had vested for Mass that he noted a penitent and paused to hear his confession. The man should have come sooner, true enough, but today — today, Father Jim never considered that.

As he raised his hand in absolution the ship hit a heavy roll. Down came the altar, the chalice, the altar cards, wine cruets and missal, ciborium and clean linens. All floated and skidded about in the water awash on the deck. The one last candle in his possession was pulverized. Father Jim's vestments were soaked with sea spray. Dear God! What a miserable "special day."

Spectators gathered; some few of the passengers must have been unbelievers for one or two laughed aloud. The little Polish lad who had volunteered to serve Father Jim's special Mass of Thanksgiving exclaimed and protested indignantly as he helped Father retrieve the articles of the altar. But he voiced his indignation in Polish and replying to his sympathetic sounds, Father Jim, sure he would not be understood, confided to this lad his secret. "Today is ————," he told him. Stooping over the wet deck in his dripping vestments, picking up his dented chalice and cracked ciborium Father Jim just had to tell *someone*.

Did the altar boy understand?

Presently he deserted Father Jim leaving him to continue

the restoration of altar and order, which, in some moments, he did. Then, ready to begin his Mass, he looked about.

Young Stanislaus was at his side, beaming from ear to ear. In accepted American hitch-hike fashion he jerked his thumb over his shoulder. It was a week-day, mind you, this "special day" — not a Sunday, but a crowd had assembled for the Holy Sacrifice. There were over five hundred. Father Jim smiled, and turning his back to the people, he began his Mass of Thanksgiving.

Among the five hundred or more were old men and women from Poland who had trod many a weary mile since 1939. As Father Jim began, they began, too — a chant which without leader or organ they sustained throughout the Mass.

Father thought of all the Masses he had offered before this; more than thirty-five hundred, in Cathedral churches and on men-of-war, in cold climes where ice formed in the cruets, in hot climes where perspiration made "tide" marks on his vestments. Grateful thoughts flooded his mind; gratitude to God for His grace, to the Blessed Mother for her protection; gratitude for loved ones and dear friends, for the prayers of his benefactors and their encouragement. And then came the climax of this Mass of Thanksgiving, — the distribution of Holy Communion to the Church universal, for many souls came forward, and all told, by their uniforms, they represented eight different countries.

It was a "special day" indeed, this tenth anniversary of Father Jim's ordination to the priesthood: a "priest forever" — to all nationalities.

MISSION TRIDIUM

NO WORK of this nature would be complete without a tribute to the Church's missionary "task forces." Here a reconnaissance of the Catholic missions is followed by the observations made by our Chaplains of their fruits. The third portion of this "tridium" comprises comments on the unsolicited co-missionary efforts of our Catholic laity under arms which has proved to be a spontaneous though little recognized apostolate.

"Task Force" Reconnaissance

Speaking in a tone of most sincere compliment, a non-Catholic soldier on Guadalcanal observed to Chaplain Thomas Reardon, the first chaplain to land on that island: "The Catholic Church is like the Standard Oil Company. It has stations wherever you go."

"At the moment of her birth the Church manifested Her international mission," wrote Christopher Dawson in the "Sword of the Spirit." "Her first public act was to speak to the nations, as represented by the polyglot multitude that had assembled at Jerusalem for the festival: men from the Middle East and North Africa and Italy. It was a kind of international broadcast that anticipated and typified the worldwide work of preaching and missionary action which even today, after nearly two thousand years, is still only in its first stages. Nevertheless wherever the Church exists there is a seed of unity and promise of peace to the nations."

Nearly two thousand years ago, as Mr. Dawson has expressed it, a small task force of twelve men set out to "preach the Gospel to every creature" — teaching those things which their Commander-in-Chief had taught them. Their *task* was to win souls for their Commander; their *force* was His word; their commission — *First* Aid to every creature.

Down through the centuries despite wars, despots, destruction, and the devil, himself, the men and women of Christ's task force have refused to bury their treasure. They have gone forth to the far corners of the earth with sealed orders — sealed with His truth — to execute His command no matter how insurmountable the obstacles. Members of this task force have suffered intense cold and heat, hunger, exposure, indifference at home and ignorance abroad which latter has more than once spelled martyrdom. And their purpose has been singular throughout; an intention to bring to all souls that infinite peace which the world can neither give nor comprehend, but without which it cannot have a lasting peace.

Father Parsons, S.J., has put it tersely: "(the missionaries) have nothing but peace in their hearts. There is no war, no force, no injustice in their Gospel. They are the invaders of foreign lands to save not to kill. They are the living proof that peace among men is possible for men of good will." General Douglas MacArthur would endorse this assertion. Following a twenty thousand mile inspection tour of the Central Pacific Area with his friend Father (Colonel) Edmund C. Sliney, Senior Army Chaplain of the Area, he stated that the United States could never repay the mission workers for what they had done in saving lives.

The successors of that first "task force" remain at their posts until death removes them. With the invasion of the Netherlands East Indies a curtain of silence was dropped over the activities of the missionaries there, but it has been reported that "not a single priest or religious among the several thousand working in the East Indies fled the islands on the invasion." When Pontianak in Dutch Borneo was bombed the government evacuated the civil population, but the Bishop, Msgr. Valenberg, and his priests remained at their "battle station." As a result, Mass continued to be broadcast each week from Batavia Cathedral and from Sourabaya, alternatively.

Pushing through the jungle islands in the Pacific our boys have learned to know and love the Catholic missionary and the natives whom he has trained so well in the ways of the Faith. Less than a hundred years ago most of this distant territory was untouched by God's "task force." Once the Vicariate of Rabaul, in the New Guinea territory formed a part of the immense Vicariate of Melanesia and Micronesia. The first missioners were the Marists, whose leader, Bishop Epalle was killed at Isabelle on December 16, 1845. Next to serve Rabaul were the Missionaries of Milan, whose last remaining priest was killed by the Woodlark natives in 1855. Next, the Society of the Missionaries of the Sacred Heart took over in 1881. But when Father Louis Couppe, M.S.C., arrived in New Britain in 1888 he found only one surviving missionary priest. Father Couppe remained for forty years and is ranked among the greatest of modern apostles.

Despite the fact there were numerous savage uprisings among the natives Father Couppe, who in 1890 had been made Vicar Apostolic, never seemed to fear for his life. He visited cannibal tribes in every section of his Vicariate and established mission stations among them: the same stations which our armed forces are beginning to know and love, today.

For fourteen years His Excellency, Most Rev. Gerard Vesters was Vicar Apostolic of Rabaul, ministering to and educating the people of New Guinea. In 1937 he was decorated for his bravery during the volcanic eruption.

"All the islands are of volcanic and coral formation," he writes. "The larger islands are crossed lengthwise by high mountain ranges, some towering to 14,000 feet, and dotted on the whole line by huge volcanoes, many of them belching smoke continuously. Earthquakes are frequent and eruptions not rare. The inhabitants belong to the Melanesian race, distributed in about seventy tribes, each with different languages and dialects. Up to recent times they were cannibals and headhunters, given to incessant inter-tribal warfare and plundering. Their religion . . . consisted of a dull fear of evil spirits which they tried to placate by witchcraft and offerings."

Later, reporting from Australia, the Bishop observed: "A great number of Missionaries of the Sacred Heart are serving as chaplains in the Army and Navy everywhere. Father Baldwin, M.S.C., was in Milne Bay, New Guinea, when the Australians staged their triumphal landing. He and Brother Fraser distinguished themselves by caring for the wounded. Father Earl, M.S.C., was with the Australian soldiers who had to defend the Owen Stanley Range in Papua. When he had to fall back with a little group, he saw that one of the wounded had been left behind. He at once returned to bring in the wounded man. But when he arrived at the spot where he was to rejoin his companions he found that they had all been machine gunned. Except for his errand of mercy, he, too, would have shared their fate. Only we who have lived in these parts can appreciate the horror of jungle warfare!"

It was on the fourth of July, 1885, that the Holy Sacrifice of the Mass was first offered on Yule Island on the shores of Papua by Father Stanislaus Verius, M.S.C. How well he and the early missionaries who followed him succeeded in their task is evidenced by the fact that at the outbreak of the present war the Papuan Mission recorded a native Catholic population of twenty-one thousand.

King Alfred the Great sent gifts to the Christians of India in the 9th century. They were the descendants of those said to have been converted by one of the first task force, the Apostle, St. Thomas. In 1943 the fourth centenary of the arrival of St. Francis Xavier on the Pearl Fishery Coast, was celebrated.

In China, July 8, 1943, was the forty-third anniversary of the martyrdom of twenty-nine missionaries who gave their lives during the Boxer uprising in 1900. Now they have been beatified. Among them was Mother Mary Hermine, a Franciscan Missionary of Mary, who with her six sisters might have escaped, Bishop Grassi urged her to flee with her sisters, disguised in Chinese clothes. But Mother Hermine replied: "We are not afraid of death nor of tortures; we came here to shed our blood if necessary. For the love of Jesus Christ, do not rob us of the palm which the Divine mercy is already holding out to us from heaven." The Bishop did not insist, so on July 8th, 1900 (not *200* or *300* A.D. but *1900*), the twenty-nine Franciscan martyrs went to their death singing the *Te Deum*.

On the other side of the world, that land so close to the scenes of the campaign of the Commander-in-Chief, Himself, the conflict between the Cross and the Crescent still continues. It has been carried on for 1300 years, for the bearer of the Crescent, Mohammed was born in the seventh century. Unlike some of the millions of lay Catholics the world over, every Moslem is a self-appointed missionary who works tirelessly to bring new adherents into the fold of the ex-camel driver. In Nigeria the task confronting the missionaries is that which would have faced them in the seventh century, when the Mohammedans swept over the northern section of Africa and hopefully buried the edifice of Catholicity already erected there. In Nigeria the Cross and the Crescent meet again in the twentieth century. Among the Hausa tribe Mohammedanism is strong and aggressive. Among the Ibos flourishes a Catholicity likened to the faith of the Irish.

But the Church of the Infinite Commander is rising again in North Africa. Eighty-four years ago the first mission stations were opened in Nigeria: for the first sixty-one years of this mission only ten percent of the priests who volunteered for it lived to be thirty years old. But their harvest now defies the zeal of the Moslems.

* * *

CORPORAL FIRST AID, TOO ...

The obscurity of time has veiled the organization of the first Red Cross society which was founded June 29, 1586. Prominent in the Church at that time was a small group of Religious, distinguished from other Regulars by a red cross on their cassocks and their long, flowing mantles. Founded by St. Camillus de Lellis, they had dedicated themselves to the sick and dying in hospitals, in prisons, on battlefields, and in many plague-stricken areas of Europe. They called themselves the "Fathers of a Happy Death": they were the advance scouts of the Red Cross, establishing military hospitals and field ambulances.

Not long after Camillus de Lellis formed his organization a pestilence swept Europe. In Rome sixty thousand died. The Camillans were welcome figures in the city and throughout the land; many of them were martyrs of charity.

After the plague was over, the Pope, Clement VIII, requested Camillus to send his brethren to aid the wounded in battle and for the first time in history a Red Cross society was on the battlefront. The Camillans furnished transportation for the wounded, set up dressing stations and military hospitals, provided medicine and bandages, and tended the sick and dying. It was difficult in those days to finance their work — there was no Annual Red Cross Roll Call — but they managed it.

Another early exponent of Red Cross work was St. Charles Borromeo, Cardinal Archbishop of Milan, organizer of the Council of Trent, and a sincere lover of the poor. When the plague struck Milan he organized relief work, personally tending the sick and burying the dead.

When our armed forces are most in need of a friend the Red Cross is at their side. So, too, in the countries occupied by our soldiers. It was the Red Cross which distributed a daily ration of milk to undernourished Arab children. But before the Red Cross arrived the Franciscan Missionaries of Mary in North Africa had been doing this same deed of mercy for over sixty years.

These are facts little known. For not too long ago a prominent magazine writer interviewed the head of one of the Catholic missionary communities of women. The writer was astonished that the doctors and nurses of this order would care for the stricken soldiers of enemy nations. "That must be something new," said the prominent magazine writer, "I thought only the Red Cross did that sort of thing."

The Superior was astonished, likewise. But she assured the writer that the first Red Cross in the world was erected on the hill of Golgotha over 1900 years ago and that from that day the faithful followers of that Cross have tried to show His love and charity to friend and foe alike. "The nationality of a soldier is no concern of ours," she told the well-known though ill-informed magazine writer, "We see in his bruised and bleeding body the Image of Christ, Himself, who died not for one race or people, but for all human beings throughout the world in every century."

* * *

AND SO, TODAY, THE RESULTS OF
CATHOLIC FIRST AID FOR SOUL AND BODY . . .

"Every man from General MacArthur down to the simplest private is eloquent on the subject of their admiration for Catholic missionaries," states Chaplain August F. Gearhard, "The natives we meet are living proof of their training and they have been of invaluable assistance to us because of the charity inculcated by the missionaries."

Chaplain Gearhard should know. A priest of the Milwaukee Archdiocese, he is a veteran of World Wars I and

II, a recipient of the Distinguished Service Cross and the Silver Star. "They aid in evacuating the wounded," he continues speaking of the natives won to Christ by our advance "task forces," "they care for fliers forced down in the jungle and assist in numerous other capacities. No returned serviceman who has viewed the mission activities in the southwest Pacific will ever have to be urged to aid the Society for the Propagation of the Faith."

"When the marines had consolidated their positions on Cape Gloucester . . . they sent patrols into the interior," reports Noel Ottaway, War Correspondent with the United States Forces on New Britain. "(There) they found the natives had fled from Japanese oppression and the violent bombardment preceding the Allied invasion. They were living in caves in semi-starvation but the men expressed an eagerness to come in to help the Americans. They made one condition — that the Marines should allow their women into the Allied perimeter as they were unwilling to leave them where they might fall into the hands of the Japanese."

These natives had been without priests for two long years, the account explains, and when they came into the American lines an Episcopalian Chaplain approached them, but when he found they were all Roman Catholics he sent word to Father Daniel F. Meehan of Montclair, New Jersey, the Catholic Chaplain of an engineer regiment, who immediately arranged to offer Mass for them.

Father Meehan reports: "Usually at home a parish priest trains two of the older boys to make the Latin responses on behalf of the congregation, but when I started the litany of the Mass these poor, down-trodden people followed it with a low chant, exactly as I have heard it in the States, although it was two years since they had attended Mass."

An Australian soldier in Papua, whose name is unknown, wrote the following tribute which appeared in an issue of the *Marist Messenger*. It epitomizes the sentiments of many: the author titles it

"FUZZY-WUZZY ANGELS"

Many a mother in Australia, when the busy day is done,
Sends a prayer to the Almighty for the keeping of her son;
Asking that an angel guide him, and bring him safely back —
Now we see those prayers are answered, on the Owen Stanley track.

Tho' they haven't any haloes, only holes slashed through the ear,
And their faces marked with tattoos and with scratch pins in their hair,
Bringing back the badly wounded, just as steady as a hearse,
Using leaves to keep the rain off, and as gentle as a nurse;
Slow and careful in bad places on the awful mountain track,
And the look upon their faces makes us think that Christ was black.

Not a move to hurt the carried, as they treat him like a saint,
It's a picture worth recording, that an artist's yet to paint.
Many a lad will see his mother, and the husbands, wee'uns and wives,
Just because the Fuzzy Wuzzies carried them to save their lives
From mortar or machine-gun fire or a chance surprise attack
To safety and the care of doctors at the bottom of the track.

May the mothers in Australia, when they offer up a prayer,
Mention these impromptu angels with the fuzzy-wuzzy hair.

* * *

"Task Force" — Chaplains

In the Pacific Theatre, this war has meant martyrdom for a number of our missionaries, but it has also brought them reinforcements in the persons not only of our Army and Navy Chaplains, but also of our American Catholic laity. The latter, by their example, have shown the natives of the Pacific Islands that Catholicism is no exclusive, isolationist ideology preached by an army of priests and nuns who have also brought them material benefits. As to the former — the Chaplains, themselves, the following paragraphs speak for their observations and experiences: they are the "Task Force" in uniform, reporting:

Chaplain Terrence P. Finnegan, from the scenes of five battles, Pearl Harbor, Midway, Guadalcanal, The Russells, and New Georgia has found a record of heroism on the part of Catholic missionaries which equals the magnificent bravery shown by the men in our armed forces. But unlike the latter, the missionaries are not "fighting to get home": there is no determinate conclusion to their campaign, not even a furlough in prospect.

When Bishop John Aubin, S.M., received his assignment to the South Solomon Islands his commission was for the duration of his natural life. To date he has seen thirty-five years of service in the South Pacific. His genial humor, his untiring zeal, and his true saintliness have won him the admiration of the natives and is now commanding the respect and affection of the men of America who are fighting other battles in the South Pacific. The Bishop's length of service is surpassed only by a Marist nun who has spent forty-five years among the natives of Guadalcanal.

Interesting as this background is, Chaplain Finnegan's associations with the men in his own division is of uppermost moment to him. He recounts them in his own words:

"The particular division to which I belonged landed in the early part of December on Guadalcanal to support the Marines who had been holding the island for four months. Our task was to drive the Japanese from the remainder of the island. After a few weeks of preparation the drive was opened on the 6th of January and completed within three weeks.

"During that time our men experienced all the horror and hardship of war, and witnessed the death of many of their comrades. They came to the realization that the only really important thing upon this earth is one's relationship to God. They found then that the material things of life which formerly had been considered so vital amount to very little. Prayer became of paramount importance. Our soldiers developed an acute appreciation of the value of spiritual well being and because of that they could perform undreamed of sacrifices and acts of heroism. . . .

"Take the case of the little Polish lad who was in the first patrol to advance against the enemy on January 6th. . . . He advanced hurriedly from tree to tree, making a complete reconnaissance of the Japanese posts. He killed three of the Japanese Imperial Marines when out of the corner of his eye he saw a projectile coming through the air: an enemy hand grenade. It struck him on the head — but it fell to the ground unexploded. This lad should have been killed but he was miraculously spared. Before the drive started he and the rest of us dedicated ourselves to the work of Mary Immaculate. She did not fail him.

"The natives of Guadalcanal appreciated the faith of our men, and their own devotion was increased by observing them. I have never seen in any land a greater love for Christ and His Church, or a more knowing appreciation and understanding of prayer than that shown by the natives of this island. One of my boys said to me, 'I thought we were doing all right as Catholics, but these natives put us to shame.'

"My first experience with the natives of Guadalcanal was at the mission of Ruavatu, opened by Bishop Aubin and later given into the care of Fathers Duhamel and Engberink. Having been shown the bodies of martyred priests and sisters I went with General Collins to the mission to make a reconnaissance of the damage done there by the Japanese and to find a fitting burial place for these martyrs.

"We arrived late Friday afternoon and went immediately to the little mission chapel to see what could be salvaged from the torn vestments and broken statues in the church. Four little native boys immediately questioned my right to enter their church. When I told them I was a priest, the first they had seen since the deaths of Fathers Duhamel and Engberink at their mission, they hurried away. The soldiers and myself continued our work at the chapel. In the distance we heard the beat of the native drums, but the sound had no significance for us.

"The next morning with a patrol we took off for the hills in the rear of the mission. Along those trails I noticed many natives hurrying toward the mission, and, as I would go by them they would get off the trail and bow. I did not know they were aware that I was a Catholic priest. About six miles back in the jungle I met two older men who indicated to me that in the bush they had hidden the Tabernacle from the mission. Having rescued it, three of us retraced our steps, for eight miles, back to the mission. We found the compound filled with men, women and children. They all told me they wanted to go to confession.

"For three hours that afternoon, with the aid of a

native catechism, I heard their confessions. That evening we had night prayers in common, and after the women had retired to a clearing a few hundred yards from the chapel, the men sat around, sang their native songs and the hymns of the Church in Latin and told me of their love and devotion to the missionaries who had given their lives for them. They told me they had been slain on the beach five miles north of Ruavatu after frightful tortures.

"Mass was to be at nine the next morning but at 8:30 the old Chief said we did not have to wait since all the natives were present and ready. Another surprise awaited me.

"When I recited the prayers at the foot of the altar the whole congregation answered me in perfect Latin. . . . After Mass we had Benediction of the Most Blessed Sacrament. They sang the *Tantum Ergo* and *O Salutaris* with a feeling and devotion I have heard only a few times before. After Benediction I had one baptism, a little native boy: I was the first white man he had ever seen and his howls gave ample evidence of his disapproval.

"These natives told me that in the South Pacific the Church is the center of their lives. Every week-end those who live in distant villages come to the mission compound on Saturday for confession, Mass and the Sacraments and do not return to their villages until late Sunday afternoons. In each village there is a little thatched chapel where they gather every night for night prayers and to sing their hymns of love and devotion . . . for myself, I could not help but notice that the Church not only takes care of their souls but that her priests and sisters are their doctors and the directors of their agricultural productions. We have often heard the missionaries tell of the need for hospitals, dispensaries and schools: now I have seen, first hand, the enormous need for such things.

"Pearl Harbor, Midway, Guadalcanal, The Russells and New Georgia will always be memories to me of courage, blood-shed, fear, and success; but they will also be an evidence of the designs of God by affording me an opportunity to see the mission territories of the Catholic Church and to observe the effective work of our missionary priests, sisters, and brothers, as well as their catechists. I thank God for the privilege of experiencing the horrors of war, for this privilege has given me happy memories not only of the faith of our grand Americans, but the extraordinary faith of the Catholic natives of this region.

"Often when I hear of our courage I am humbled by the courage of the Sisters and priests whom I have met and who, without any of the conveniences afforded by the United States Army, left home forever to find another home among the natives in the South Seas. I shall never tire of speaking about them and boosting the cause which is nearest their hearts."

It was on Guadalcanal that a Dutch Catholic missionary stuck to his post throughout the Japanese occupation and the battle which followed, rather than desert his charges. When the first rescue ship moved into isolated Beaufort Bay after the American victory, Father Emery deKlerck, a Marist Missionary, was found ministering to his native converts as serenely as if he had been unaware that he was living on the edge of a great battlefield. Undaunted by the murder of two fellow priests and two nuns by Japanese soldiers, the missionary continued at his post and evaded evacuation. The Netherlands awarded Father deKlerck the Bronze Cross for his "courage and intrepidity against the enemy . . ."

For a time the Japanese had behaved relatively well toward the Missionaries. Coming into Beaufort Bay, the

Japanese patrols would politely draw up in a line and present arms to the priest, and an officer would salute smartly and ask, "Please, have you any arms concealed?" But as the Americans pushed the Japanese off the island their "manners" preceeded them into the sea.

"I'm going to tell you a different side of this war," wrote Captain Eugene Schoenfelder, a U. S. Marine, to Brother Aubert of Manhattan College, N. Y., "a side that is important to us as Catholics."

"Our Chaplains, Fathers Fitzgerald, Reardon, and Gehring — what men! Thank God the Navy saw fit to give these sailors to the Marines. They are a source of pride to me. They make me proud that I, too, am a Catholic. Everyone is inspired by the courage, faith, and devotion to the men that these priests have displayed. There have been many occasions when the men sorely needed them: they never failed.

"It's Sunday down here. This morning Mass was offered by a local missionary, Father Wall, who has been here for ten years. Until just a few days ago he had been hiding in the jungle. He, and several others, had been rescued by a Marine patrol. There is still a nun up in the hills who will be brought down as soon as she is well enough to move. She is in very poor condition and probably very old as she has been on the island for thirty-three years.

"Just as important as our reconquering the island, Father Wall told us, was our attendance at Mass and communion and our visits paid to the Padre. Before we came the natives sometimes wondered if their religion was practiced by people 'in the great world.' When we came from the large, far-away United States and attended Mass as they, themselves did, they knew our Church was catholic. Our actions did much, indirectly, to strengthen their faith.

"Father Wall said the news would spread throughout the

tribes and islands and would be talked of for months. Marines, missionaries! Helping to conquer the spiritual as well as the mundane world. Brother Aubert, impress this upon the boys before they leave for the services: — *Catholics are now being met by many people for the first time. Here is an opportunity to break down prejudice. We can show what we really are. And we don't have to preach — we merely have to act — properly!*"

Chaplain Francis Gorman of Chicago, writing to the Society for the Propagation of the Faith under date of May 11, 1943, reported that: "The foreign missions have won many new benefactors. After this war the Propagation of the Faith should have little trouble raising funds for this particular charity.

"Few of us living in the lap of luxury ever realized the sacrifices that our Catholic missioners were making out here in the Pacific. The only friend these poor natives have is the missionary; and these same natives would shame most of us in the way they live their Catholicity. They are not just fair weather Catholics nor just Sunday worshipers. They have a far better knowledge of the liturgy and the Mass than many of our metropolitan Catholics back home.

"You should hear them chant the Mass; you should hear them pray in Latin. I never realized the value of universal Latin until I had occasion to offer Mass for natives whose tongue I could not speak. Because of Latin, I could pray with them. The prayers after Mass which we are accustomed to recite in the vernacular they always say in Latin. In each village they recite every evening the Rosary in perfect Latin."

An Army officer, veteran of the Burma campaign, writes glowingly of Father Jeremiah Kelleher, missionary of St. Columban, who served as chaplain with the British units defending Burma in the spring of '42.

Prior to the fall of Rangoon, during the bitter struggle

of the Salween River, Father Kelleher was constantly in the front lines, ministering to the wounded among his own Kachins and other British units. He stanchly refused evacuation as long as any troops remained in the area.

It is noted, too, that the presence of Catholic Sisters acting as doctors and nurses in India is making the people of that vast country medical minded — to their lasting benefit. In Pindi the Medical Mission Sisters desire very much to build a hospital. American soldiers have visited the Catholic Medical Mission Sisters in Pindi and also Dacca — and have left their generous contributions toward more hospitals and medical service for the Indians.

India presents special problems to missionary and Chaplain. There is for example, the language barrier. Chaplains serving in this field would have to be able to converse with native troops in Marathy, Hindi, Malayalam, Tamil, Kanaresse, and in Telugu, just to mention a few. Here the Chaplain readily perceives the need of a native clergy.

In Libya ten Oblate Fathers are acting as Chaplains for the native and English soldiers. Basutoland is practically an independent nation but 58,000 of her men volunteered to fight for the Allied cause. Of this number 22,000 are Catholics and 36,000 pagans. Hundreds of catechists formerly under the supervision of the Most Rev. Joseph Bonhomme, Oblate Bishop of Basutoland, South Africa, have joined the armed forces where they continue their work between military engagements. Whenever possible they assemble the men for morning and night prayers. They hold instruction periods and have been responsible — these lay missioners — for hundreds of pagan baptisms. Paradoxically, says their Bishop, it seems that through the horrors of war the seed of faith is being widely sown in Africa by Africans.

Rev. Harold W. Rigney, Chaplain of the Army Air Corps has an observation, too:

"Let me tell you the Armed Forces of the United States owe a great debt of gratitude to the Catholic missionaries of Africa. For over a year the Catholics of the U.S.A.F.I.C.A. were cared for only by missionaries. The fact should make the American Catholics more appreciative of foreign missions and encourage them to continue their generous support of them. . . . "

So the war brings new roles to our Catholic missionaries: being versatile souls accustomed to abandonment to the will of God, they carry on wherever that will directs them. Thus, from Africa, the White Father, Rev. F. Gaffney, is acting as Senior Chaplain in Iceland, while a Canadian member of the same order holds the same post in Ceylon. Another, Father Haskew, assigned as Chaplain to the native troops in Africa, reports:

"It is just six months since I was commissioned . . . the job is not nearly so good as being on trek in a mission! . . . Language is a handicap, for the natives are a mixture of all the tribes of West Africa. They know a sprinkling of English but it is completely inadequate for things spiritual. I find a good percentage of Catholic Europeans, and these I have been able to help. But, in spite of difficulties, my mere presence in the Brigade achieves the main object for which I am here, namely to preserve the Faith in the hearts of those where it is a seedling, a plant, a rooted tree."

At this point, if these pages have been attentively read, there should be uttered a resolve never again to protest: "Why support foreign missions when so much needs to be done at home?" And while so resolving it might be noted that the average American Catholic contributes 7 cents a year to foreign missions, reserving only 1 cent a year for home missions.

Our Catholic laity under arms is resolving to change this picture.

Master Sgt. Wayne L. Smith had rather definite ideas

on the subject of foreign missions, but he has changed his mind. "When we used to talk about foreign missions at home," he writes, "I used to think it was a waste of money, but since I've seen the work of the missionaries I know better."

Sgt. Robert McNamara, an air gunner and winner of the Silver Star, was forced down in the jungle. "After three weeks I was found by an English speaking missionary, Father Vic. Were it not for his kindness I would not be alive today. Please say a prayer for this missionary and ask my friends back home never to pass up a collection for the foreign missions. They are a great work."

Chaplain Thomas A. Duross, S.J., informed the Society for the Propagation of the Faith: "I have been working a little project here in Africa . . . perhaps other priests will put it into effect, too. With the permission of the Army I have taken up a collection every Sunday which is divided between the local Catholic Mission and the Chaplain's Fund for the upkeep of the chapel, charity and welfare needs. In December and January the local missions here received $600 as a result, and $200 worth of supplies have been sent to the French missions. Remember us in all your prayers."

The busy partner of "Padre Inc.," Father John F. Culliton, enclosed a check for $1,053.85 from the boys of his division when reporting to the Propagation of the Faith. "My boys want these funds to be ear-marked for the Solomon Islands Missions." Chaplain Edward Whelly collected $600 at one time for the same purpose. The soldiers, sailors and marines who fought on Guadalcanal donated $825 in one collection for the missions: more, a Protestant officer gave twenty dollars for Catholic Missions with the comment, "I never had any use for missionaries, I thought they were never any good. But those I have met are the bravest men I have ever seen."

The average donation of each enlisted man has been

about five dollars and well over one hundred thousand dollars has been given by our men who have seen the missionaries, their converts, and the hardships under which they have labored in the South Pacific.

At the call of the President of the United States Americans annually observe a day known as Thanksgiving. Large sums are spent for edibles to celebrate this day. At the call of the Holy Father Mission Sunday is observed each year, the next to the last Sunday in October. Could each of us donate the price of a Thanksgiving banquet in thanksgiving for our missionaries? Yes, there is always the landlord or the mortgage interest, taxes or tuitions — and generally a turkey on Thanksgiving Day. How about turkey for the missionaries?

Unsolicited, the missionary and the Chaplain "task forces" have found rising to their assistance a voluntary apostolate among the Catholic laity under arms. War has its blessings: perhaps the greatest is that its dangers, separations and uncertainties turn mens' thoughts to their beginning and their end. Souls adrift seek an anchor to windward. The Catholic laity under arms has been ready to show storm-tossed and fog-bound souls the only safe anchorage. Neither missionaries nor Chaplains by vocation or circumstance, they have aided many such to acquire that peace which the world cannot understand, and therefore generally shuns. The following section describes some of the work of this spontaneous apostolate which could only exist among a people accustomed to live in a land where free speech and freedom of religion prevails.

"Task Force" — Catholic Laity Under Arms

It must be kept in mind that the Army and Navy have not commissioned Catholic Chaplains to serve Catholic personnel *exclusively*. Whatever the Chaplain's faith, be he Catholic, Protestant, or Jew, he is obliged to serve all men under his charge. Further, when a superior officer is faced by a serviceman who has a personal problem which the officer finds "beyond him" he generally evades the issue by advising the questioning man to "tell it to the Chaplain."

This is how Jim, of a stictly chemicobiological philosophy, brought his problem to a Catholic Chaplain, asking, "May I speak to you, Sir? I am not a Catholic." Instantly the Chaplain sensed Jim's source of trouble was that he had no spiritual stabilizing element in his life. It is hard for any Chaplain to reach such a man to help him through a tough spell: a fact Jim's superior officer no doubt recognized when he gave him the "tell it to the Chaplain" brush-off. But the officer should have recalled that there is little time in the daily routine of a service Chaplain to rebuild a philosophy of life. This task should have been begun, and on sound verities, at a mother's knee. But alas, mistakenly, many parents want their children to grow up with an "open mind," as they suppose their own to be, on the subject of religion. Mr. Chesterton once said: "Merely having an open mind is nothing. The object of opening the mind, as of opening the mouth, is to shut it again on something solid."

Fortunately for our Chaplains in the Armed Forces some years before the war, a modern application of St. Paul's epistolary technique was begun at Kenrick Seminary, Webster Groves, Missouri. It was devised by Father Lester J. Fallon, whose motor-mission activities brought so many questions on the basic articles of the Catholic Faith that he enlisted the aid of the seminarians in answering them by correspondence. There is an authoritative answer to every question. Replies from an informed Catholic source would not vary whether they came by correspondence or word of mouth. If the source be Catholic, it is not important where or of whom the non-Catholic's questions are asked; a correct Catholic reply *is* important.

Since Pearl Harbor questions about the Catholic Faith have increased enormously, as thousands for the first time have met Catholics and witnessed Catholic services on board ship, in jungle chapel or foxhole sanctuary. So, the correspondence courses which were begun in peacetime have been expanded to meet the demands upon them. They have saved the Catholic Chaplain much labor, for often before a question reaches him it has been asked of a Catholic layman under arms who has, in turn, solicited the aid of the correspondence course.

"I don't know about you fellows," said a new arrival at an Army post one day. He had been transferred from another camp. "But it took a war to make a Catholic out of me," he continued.

He was one of a group of seven enlisted men. His experience was true of them all. Like thousands of others in the services they had, for the first time in their lives, come to realize the emptiness of men without God. They had cast about and found that men who understood the value of prayer were able to adjust themselves to war. Naturally of such men they had asked questions.

Writing of Catholic men under arms, Father Stephen B. Early, S.J., has said: "The practice of religion in the Armed Forces immeasurably surpasses that in civilian life." "You want to know about the conduct and the religious life on

Guadalcanal," writes Chaplain Kenneth B. Stack. "There, more than anywhere else, one realized that a father and mother, being closest to them, should be an example for their children. No kid hits any higher than he aims. If a father and mother haven't given good example, one can expect no better of their son. The effects of the life we live nowadays . . . were apparent on Guadalcanal." Commenting on the general disregard for law and the authority which comes from God, Chaplain Stack continues, "With few exceptions, a man will be just as good in the Army as he was in civilian life. . . . The Army neither makes nor breaks a man. His character was molded before he got into the Army, and he is only as good as his father and mother have made him."

Perhaps this indirect approach will react upon the reader as cryptic and inconclusive. But the Chaplain does not leave the subject at this point. He enlarges slightly: "All the Catholic boys who attended Mass regularly at home were equally anxious to attend whenever possible in the Solomons. . . . Religion with the soldier is an intangible thing. . . . To me the real Catholic . . . is the one who goes along quietly, minding his own business, but just as quietly fulfilling all his religious obligations and conducting his entire life on the principles of Christ."

A cadet in a California flying school noticed the broad opportunity for Catholic Action open to any layman. As he knelt beside his bunk to pray at night he was surprised the scoffers were so few. In a short time other cadets asked him what his religious beliefs were. The first one to approach him frankly said he was seeking better answers than he could devise for the problems of life. At the end of three weeks, this first questioner was baptized.

Nor did this apostolic cadet merely wait to be asked questions. He was quick to discover the lax and fallen-away Catholics in his outfit. He found, surprisingly, that the latter had to be re-convinced of their need of religion.

Private nightly prayers at the bunkside is what one Catholic Army Air Force lieutenant calls "selling religion by displaying samples." It was his experience that many men had been sent in search of religious guidance merely from witnessing this practice. These men will not go to the Catholic Chaplain for answers to their questions. That would be "getting too curious" or "carrying a thing too far." Instead they go direct to their fellow Catholic servicemen, commissioned or non-commissioned.

Perhaps it may have been this lieutenant's "sample display" or one like it which attracted a fellow officer who began instructions at Fort McClellan, Alabama, and completed them in Bougainville. Jack M. Tucker, Army Public Relations officer, tells his story:

"The first thing that Father Joseph B. Delahunt, a priest of the diocese of Syracuse, did upon arrival at a South Pacific base was to look up a Lieutenant from a small town in New York State, and for a good reason. Father Delahunt had been giving him instructions in the Catholic Faith when both were stationed at Fort McClellan, Alabama, and the Lieutenant was moved out before the instructions were completed. Shortly after their meeting on the other side of the world, the Lieutenant was received into the Catholic Faith." Father Delahunt in Bougainville, found himself just where he wanted to be, in the muck, intense jungle heat, and insect-infested hot-spot which is that Island. "I'd hate to come out of this war," he told Lieutenant Tucker, "if I do come out, without seeing actual combat. By combat I mean administering to men under fire in the front lines."

"I'd hate to come out of this war," he told Lieutenant Tucker, "if I do come out, without seeing actual combat. By combat I mean administering to men under fire in the front lines."

Father Early reports other "sample displays":

"Over in North Africa a company of Army Engineers

doing repair work in their own shops, were too few to rate a Chaplain. But every night in the barracks with the C.O. leading them, they said the rosary in common. They weren't panty-waists. They weren't pietistic past-presidents of the Holy Joe Confraternity. They were the hardest, cleanest, highest type of American Catholic men.

"Up at Fort Lewis in Washington, a young rookie in the barracks on the first night knelt down by his bunk, and said his rosary. At first there was a small commotion — but he was well over six feet and 190 pounds. About a week later a lad by the name of O'Brien saw a lad by the name of Murphy get down on his knees.

" 'Well,' he said to himself, 'if a Murphy can get on his knees so can an O'Brien.'

"Before the month was out, every Catholic in the barracks said a nightly rosary, while non-Catholics maintained a reverent silence. Nobody laughed."

Because many "sample displays" on the part of our Catholic laity under arms have been fruitful, let the reader not assume that there are no Catholic servicemen in the Indifferentist camp. Our Chaplains have discovered about one third of the Catholic men under arms do not practice their Faith: they are not 100 per cent apostles by any means. The good that one apostolic soldier may do, can be undone many times over by one lax or fallen-away Catholic.

These are many. For example, over a period of twelve days one hundred sixty-four Catholic men were admitted to an Army Hospital. The Catholic Chaplain discovered sixteen with bad marriages, eight who would have nothing to do with him, twenty-six who had been away from the Sacraments for from eighteen months to eighteen years and four who had not received their First Communion. From the southwest Pacific a Catholic Chaplain has observed: "I'm convinced the glowing accounts of the religious revival in the Army are unadulterated nonsense. God help the Church if the faith manifested over here in the Army represents the highest degree of religious fervor among Catholics."

In many minds the thought will arise, "How can the men, especially the Catholics, remain indifferent when danger and death filter the very air they breathe?" Forgetting that danger and death hover every bath tub, every flight of stairs, every automobile, bus, or train, the questioner may continue, "Indifference is easily understood at home where everything is safe, secure, and pleasant; but on the battle front?"

In truth, reports are conflicting. But from many sources culled to date, no dramatic account of re-conversion at the point of bayonet or block buster has come to light. And why should it? Catholicism is a Faith which appeals to man's reason. He must think before he lives it: bombs and bayonets allow no time for meditation. Chaplain Gearhard has been reported as saying that while it is quite true there are no atheists in foxholes, it must still be remembered that a man's entire time is not spent in such close quarters.

Yet Chaplain Reedy, dropping in on a group of soldiers in Alaska, found them seriously discussing ways and means to retrieve fallen-aways. Chaplain Gaffney, reporting from his post on a large battleship, has stated: "Catholicism in the fleet is tops . . . in spite of conditions which are nothing short of pagan. On Sundays I have two Masses and the attendance accounts for over eighty-five per cent of the men, and many of those not present are on watch. There is a daily Mass on board at which some fifty to seventy-five men are present each day."

Father Early holds that the fighting men have come to realize what Holy Communion really means. Canadians crossing the Channel in their commando raid on Dieppe — which was expanded to a mass assault on D-Day — attended Mass, offered by their heroic Chaplain Sabourin, and received Holy Communion. One of them wrote: "On the water without lights and in the tense silence, one thinks of a lot of things, and it was strange how my thoughts stayed with that Communion. They say any man who claims he is not afraid of going into action is a liar. The truth is, I had been afraid for awhile, but I was not afraid on the boat. Holy Communion had made me feel there was nothing to be afraid about."

Chaplain Ralph M. O'Neill, S.J., when serving with the 35th Air Service Group, reported in the *Jesuit Seminary News:*

"The attendance at Mass has been very gratifying and the numbers who receive Holy Communion each day are inspiring. . . . I was the only priest on board. . . . On each Sunday I offered Mass three times with an attendance averaging about five hundred a Sunday. . . . All could receive each day as Viaticum — therefore with no need of any fast whatever. . . . On the last evening of May we had the Rosary on the upper deck in the open with officers, nurses, enlisted men, some Navy gun crew, civilian sailors, all participating. . . . One day a crowd came to my room led by an Italian from Chicago and his pal, an Irishman from Cincinnati. They wanted to have a Choir and High Mass. I said no can do that. No music, no books, no even what on board. But we typed out all the words of the Kyrie, the (*cut from the reverse side by Censor*) est from the Missal. We (*cut by Censor*) Mexican from Texas who could hum the 'Missa de Angelis.' We had a Canadian Frenchman from New England who had had some training and could write down the music as he heard it. So he wrote the notes which the Spaniard hummed and armed with that homemade edition taught the rest of the crowd in the Choir. We had High Mass in grand style on the last two Sundays on board. . . . Keep after our Blessed Lady for all of us. . . ."

It cannot be denied that opinion varies. From the same sector where one Chaplain reported faith among the Catholics at a low ebb, Chaplain John E. Leonard reports:

"In the days to come the Catholic Church will be proud of the American men now serving their country. In battle they have found their Faith and have come to a real appreciation of the Real Presence." On the Feast of Corpus Christi, neighboring military units stationed in the South Pacific were astonished to see one thousand troops parade by on an unscheduled march of many miles, to attend the one Mass to be held that day. It was held in a coconut grove cathedral, the altar a mahagony log. The gorgeous display of innumerable tropical flowers had been brought from other islands by the boat load. And during the "watches" before the Blessed Sacrament which was exposed the night long, frequently the men had to leave their posts immediately before the crude altar and "watch" from nearby foxholes.

A few paragraphs from *America* contrasts with these comments: "The spirit isn't always good. A junior officer in a combat group in Alaska gave one Chaplain lots of difficulty. He had an Irish name and a Catholic education and a sharp tongue and a black heart. It was Christmas Eve . . . (and) after a struggle that almost cost him his life, the Chaplain arrived at the camp and on crates of vegetables dragged in from the kitchen, set up his Mass kit. . . . Everybody in the camp was at the Mass, even he of the black heart who stood in the back, and didn't kneel at the Consecration. Then came Communion time and a hundred and fifty moved up slowly to receive. . . . Slowly in the back of the barracks the spirit of the men's Catholicism penetrated a cold heart. Before the end of the Communion he was on his knees.

". . . the Chaplain moved on. Thirty miles down the road another group waited for their Christmas Mass. Confession before Mass is the usual custom. . . . again (at Communion) every Catholic received — finally there was one man left.

He lifted his face slowly, the tears freezing on his cheeks. The black heart was black no more.

"He didn't attend Midnight Mass this year. He died the death of a hero at Attu."

It was not the Babe of Bethlehem directly Who helped the Chaplain assigned to a western Coast Defense detachment. Again lifting from *America* "... if the truth must be known the Catholicism of this outfit was nothing to write home about. The Padre worked hard. He offered his 16:30 Mass in a different group (of the twenty or more) each day, for a faithful ten or twelve out of a possible hundred men. The Spirit of Catholicism in this Coast Defense outfit was dead, and all it needed was decent burial.

"The solution of your difficulties is the Holy Hour," advised a kindly prelate.

" 'The Holy Hour!' exclaimed the Padre, 'Monsignor I can't even get them to Mass.' "

" 'Try it,' replied the Monsignor gently, 'you'll be surprised.' "

The Padre tried it.

" 'This is no gimme-prayer,' he told the *seven* men who showed up for the first Holy Hour. 'We'll offer an hour of reparation for the sins of the regiment. The whole idea is to atone to God for the insults the regiment gives Him.'

"Soon fourteen separated groups of the regiment held a monthly Holy Hour. They made them whenever they could — some of them in closets, one in a barber-shop, one outfit was even willing to make it in the latrine building, the only lighted place in the camp. . . . Now ninety per cent of the Catholics in that Coast Defense outfit attend Mass regularly. The Padre never comes into a new group without finding some lad ready for confession — the first in five, even ten years. An officer recently praised the Padre for the fine spirit of his men, for their regularity at religious exercises.

" 'Me?' the Chaplain said. 'No, sir. We owe the Catholicism of the regiment to ninety-eight men who spend their hour each month praying for us.' "

Add to the foregoing, if you would believe there *is* religion in our armed forces, the unpublished request for a ciborium from a chaplain in one of our southern states: "Communions are on the increase and it is becoming very inconvenient to consecrate large numbers of hosts without a ciborium. The smallest and the oldest would be received with a grateful heart."

And from British Guiana, a soldier magazine called *Tropical Daze* editorializes: "Will we ignore the principal lesson of the war, which clearly points out that a nation and its people cannot forget God and still expect to be showered with His blessings or hope for a shield of divine armor to protect them from the thrusts of an aggressive army? Pray God that we do not, for only with a nation which realizes its place in world affairs, a people who understand their responsibilities to their fellowman and to God, serious minded men and women anxious to serve their God and their country in peace as well as in time of war, will we ever attain the peace and happiness for which we are now fighting."

In Italy a Polish priest, Father Stanislaus Targosz, was asked by two American units who had no Catholic chaplains, to offer Mass. "It was impossible to go on Sunday, but I went Monday evening. When the servicemen learned that I could hear Confessions in English I got plenty of work. It was moonlight. The altar stood at the entrance of a tent. Before the tent were about fifty men, holding rosaries in their hands. Many of them received Holy Communion. The next Sunday I went to them again and offered Mass in a barn which had been cleaned, supplied with plenty of boards for pews and thus converted into a chapel."

Contrary opinion among our Chaplains notwithstanding, it is not possible that the "sample displays" of practicing

Catholics should fail to make a salutary impression upon the non-Catholic serviceman who "never had any religion." Conversion is not an act of Chaplain or layman: it is an act of God, alone; good example is one of His tools.

"Our boys away from home are ardent apostles," asserted Father John P. McGuire (Senior Partner of "Padre, Inc.") when he was home on furlough. "They attend services faithfully, not only on Sunday but morning and evening as well. They impress non-Catholics with their sustained devotion to the Catholic Church . . . the success of our missionaries has been aided considerably by our Catholic boys . . . it is consoling and heartening to learn how seriously our troops take their religious duties."

That the war, which God permits, gives impetus to conversions cannot be denied. These totaled over ninety thousand among Americans in 1943. Undoubtedly this record was made possible partially by the Catholic Action of the newest, voluntary "task force" — the Catholic laity under arms. It is sometimes forgotten by the laity that ours was a mission country prior to 1908; that while in 120 years the population of the United States increased 24 times, that of the Catholic Church in the United States increased 600 times. In the face of these figures, one is almost disappointed that the Catholic laity under arms, and in civil life, too, is not doing a better task! In 1943 there were one or more vacant chairs at many a rationed dinner table of a Catholic family. But the "home folks," reading these pages, may draw consolation from the fact that mayhap their absentees are among the laity in uniform who at the same time, of their own volition, are also "out on the missions." If God calls them to their eternal home, these laymen will not go to Him empty handed.

WAR IS MY PARISH

Because general background and many incidents are identical in the activities of many Chaplains, these three stories, except where names are specified, are based on the experiences of

    Chaplains George Flanigen
               John E. Leonard
               Edward Bradley
               Harold J. Barr
               John Culliton
               John Powers
               Joseph Patrick Mannion.

" . . . Upon a Midnight Clear." — (?)

APO San Francisco, to you;
Somewhere in the South Pacific
to me — December 26th

My darling: —

What I dreaded is over, — my first Christmas away from home and *you*. — Are your eyes still as blue, Mary, and your — oh, well!

I know what your Christmas was like — our Christmas mail has not come in here yet — but you missed me, I hope, I *hope!* But you put on a smile, I'll bet, and you probably went to see mom and dad, maybe you three even talked about me, — and who knows what you might have said? Maybe you had turkey, — I had recollections of turkey: I understand ours is held up with our mail, maybe for New Year's we'll get it. I suppose you had a tree and tinsel and lights; maybe snow outside, crisp, snappy air, anyway. What I could give for one sniff of air like that! And of course you had midnight Mass, and friends to see and a lot of presents being handed back and forth, — jokes and laughing, too. Your Christmas must'ave been somethin' like that, because that was the way I was thinking of it — when I was. A few days before, I must admit, even in spite of the good work our Padre was up to, I was feeling darn sorry for myself. But not for long. Our Chaplain, Father Jim — a swell guy — wouldn't give us a chance to be homesick — much. I'm telling you, Mary, I'll never forget this Christmas — *never*. Away from home? YES! And from you? Double YES! Homesick? Ask me another, baby. But there were *compen-*

sashuns, as Private Purkey would spell it. We had midnight Mass, too. Here's the low down.

About two weeks ago Father Jim began planning for it. He knew the whole outfit was down in the mouth — seemed like even my bones ached homesickness, and I guess I was not the only one. This jungle heat is no inspiration to pull out of the dumps. The least exertion lays you low. I'm dripping now, just writing to you, and I'm sitting on what is — or should be — a cool hilltop. I can see water all around me, but it don't look cool, the blazing sun is on it — it's always a blazing sun, here. The uncomplimentary gag "You drip!" is no go here — everyone is dripping all the time. It's especially hard because most of the fellows in this outfit are from cool climates and I can tell you we all needed at least a nip in the air to get a sniff of Christmas spirit. There's no nips of that description in this hot spot. It's an effort to get about even on our "appointed tasks," and they're nothing as soft as golf. Well so, Father Jim laid out some more exertion for us and on our free time, too. We don't get no time off for extras like that. Of course if we'd of been in some spot where we could do Christmas shopping I suppose we wouldn't of called it exertion. Sorry, baby, I couldn't do up a fancy gift package and send it home to you with love and kisses — but how did you like what I did get sent to you?

Well like the heat, bugs, mud, and regular routine was not enough, Father Jim lets fall the notion of a special chapel for midnight Mass "to accommodate the crowds," he said. Here, it's just like home about crowds on Christmas: most everyone turns out for church call that day. But being our first Christmas we kind of thought Father Jim was spreading it thick; none of us thought it would happen here: but we lived to learn the Padre had his bearings right.

Father Jim is one of those cheerful souls who is genuine so his cheerfulness doesn't gripe you — it kind of shames you. He never takes a day off and he never barks due to the heat and overwork. You might guess he was Irish and you'd be right. Seems to be always bouncing about on the double: I guess he has the secret of perpetual youth and believe it or not he got us pepped up spiritually and physically. He had us rig a big hospital tent for the Christmas Mass. He swore our camp chapel wouldn't be half big enough. Then on account of when it unexpectedly rains here, which is almost every day, it's never a two cent drizzle, we pitched in and dug a drainage ditch all around the tent — just in case. This turned out to be a good hunch, only we didn't dig it deep enough. Mind you, Mary, we get no time off for this sort of stuff and no extra pay. We have to do it on our own.

Seems like the tent and the digging got some of the indifferent Catholic fellows curious — there's still some of them around. Some gave us the razz for being a bunch of suckers, but Tim McCarthy fixed them. You remember Tim — the red-head I brought home for dinner my last leave. Tim turns the heat of his Irish temper on these hecklers — and in the tropics the Irish temper gets extra special hot, and the scoffers backed down. Then Tim hands them shovels, too, and the Padre wears a broad grin. Heck, some of the Protestant and Jewish guys came along and volunteered to help. We had to build an altar, of course, which was no work of beauty, but still an altar. We set it on a platform about half a foot from the damp ground. Then we hung a bed sheet behind it and one of the fellows made a big cross. We polished the vases for the flowers till they glistened something wonderful. You wouldn't see a better polishing job in a cathedral. And wait till I tell you about the flowers. In case you're curious, the vases was seventy-five millimeter shell cases — and okeh, too. We hauled in iron bomb fin cases for the pews. One of these the Padre had us set off to one side for the confessional — the kneeler was the dirt floor.

We are still at this job of chapel building when the

Padre says to us one nite, "Now look here, fellows, I don't need to tell you Christmas is coming and there's no stores around here open for gift shopping. But that's no reason why you all can't send a present back to your folks." We looked around at each other, and some of us wondered if the heat had got the Padre at last. "Here's how," we heard him say, "I've got a good stock of gifts on hand, all you need to do is pony up — not cash, but some more of your free time —." To read this Mary, you'd think we was all off here on a holiday. Our free time, I can tell you, is not enough for a college education, not even one of those accelerated courses we read about. "Extra time," says the Padre, "for the next nine days. This way you guys can make a novena for someone you love back there . . . " guess you have caught on by now, sweetheart. I didn't need no second push to make a novena for *you* — maybe for *us* and our marriage when I get back, huh?

Well, Father Jim lays in a stock of V-Mail letters — like the one you must'ave got by now — and he says he will send one to every person back home that a novena is made for. He signed them all, as you probably notice by the one you got. The catch was we all went to Mass and Communion for nine days ending up at midnight Mass, Christmas.

So it was like this that Father Jim kept us thinkin' of the spiritual meanin' of Christmas. No jim-crack junk to buy for presents, — only God's love and protection for us all and our folks to ask for. Honest, baby, this made me feel awful close to you — like as if what we believe is the only real thing that keeps us together and going to get back home. I kept hoping maybe you were making a novena, too, but I suppose your pastor didn't have any such bright idea: you don't think so well, sometimes, when you have a good, soft bed every night, — only you sleep well, that's a cinch. *Did* you make a novena, too?

Well, so we build the chapel and make the novena and Father Jim figures still we might be homesick so he gives out he wants a choir for the midnight Mass. So my beautiful barytone goes on maneuvers for that, too; a little rough going in spots, I might say. We had to have a lot of training, and we discovered there ain't a opera singer in the outfit.

Like this, the Padre had everybody working on something. He is a smart one. One of the fellows is artistic, so the Padre let him paint a picture of the Blessed Mother and Child that's a knock-out: it could be on the dome of St. Peter's an you couldn't tell was it done by a Yank or that old Angelo fellow. Carpenters, electricians, engineers — officers and men — everybody was working, never mind did they all believe as the Padre did or not, they worked.

About a week before Christmas the Padre gets a package from the States. Some Council of Catholic Women somewheres has sent him, by his request we found out, a new set of gold vestments and altar linens which he naturally holds back for the big event.

I'm telling you, Mary, my darling, how Father Jim held out I don't know. He heard thousands of confessions sitting on his bomb fin confessional. He gave out he wanted every baptized Catholic to show up and get his conscience on the dotted line. You won't believe this, but even the hard-shelled crabs came out before our novena was over. There's something about Father Jim's personality that cracks the hardest nuts.

He says to us one nite at Mass: "You fellows are going to have a Christmas like which we have in the seminaries. No commercial stuff. No cash and carry shopping. No bargain counter crushes. No traffic jams and your arms loaded with bundles. None of that racket. This Christmas is going to be the real thing. No worldly diversions this year to make you forget what happened on the first Christmas and how important it is to every last living one of you. Now, in the seminaries," he goes on, "the Spirit of Christmas crops up and blooms like a rose, because everyone in the outfit takes full part in the Mass and goes to Communion. That's how we're going to have it here."

Well, Mary, so came Christmas eve — just twenty-four hours ago as I sit here writing about it to you. It was like now, a beautiful tropical night. No snow. Mild and warm and a million bright stars twinkling down on us; the Padre says most likely the same kind of climate as at Bethlehem. There was crowds, but they weren't tired and cross last minute shoppers looking daggers at you for stepping on their sore feet. They was fightin' men, all streaming towards our special chapel. They came in jeeps and trucks and on foot. We all knew we wasn't home, but somehow it seemed almost okeh, Mary — like we *was* where He wanted us to be. I guess that was the Christmas Spirit the Padre had been working for.

But wait: as I just stopped to smack a bug, I remember I didn't tell you about the flowers. P-o-n-c-i-a-n-a blooms, Father Jim called them. They might as well have been orchids, except they are brilliant red. Here's how we got them — none of your dainty snipping with a pair of scissors. We take a jeep, a hatchet, and a bayonet and we drive into the jungle with the Padre. These trees grow like weeds splashing all the jungle foliage with buckets of red. One of us would climb a tree, hack off a branch with the hatchet and the fellow on the ground grabs the fallen branch and strips the blooms off with the bayonet. When we had a jeep full of flowers we went back. Those brilliant red blooms against the white bedsheet backdrop were a sight to see. They kinda shriveled in the heat, but they held their color. Or maybe it was our drinking water in the vases they didn't take to. It's got so much chlorine in it we don't take to it much ourselves.

About eight, a pal of the Padre from another base came

and the two chaplains heard confessions until it was time for Mass. I forgot to mention we'd put up this chapel near where about three hundred of our buddies are buried. That was our own idea, which Father Jim okehed, — seemed like then we could get them in on it, too. Well, so Mass starts off fine — you never heard such a choir; even with me in it, it was good. Before we get really under way, though, the Padre takes time to baptize 34 fellows who had taken instructions; he's got as many more coming, too, but he couldn't finish with them so's to get them under the wire for last night. You know how particular the Church is about these converted fellows knowing what it is they got converted to.

So Mass begins, we sing and everything is good and spiritual. At the consecration the choir stops singing, and off on a hill not more'n a stone's throw away, soft and faint, a bugler sounds taps — that get's our fallen friends in with us, too. Everyone in the chapel, — plenty of officers and men not Catholics, too, goes on their knees to a man. The choir couldn't've sung if it had wanted to just then.

But that was the last of our beautiful choir anyway. Old man nature couldn't hold off for another hour or so — no, then and there at that moment, he must start emptying all the lakes of the world over the tent where we was. You've got to see one of these tropical rainstorms to believe it. Rain can come up sudden at home, like one big giant's sneeze, but in the tropics it's not only sudden but abundant. The old man holds back nothin'. Our drainage ditch was filled in a wink and the water begins oozing under the tent flaps — no, oozing ain't the word. 'Rushing' is more like it. This was the worst storm since we were stationed here. The altar's platform was under water in no time — when the Padre genuflected he came up with a muddy knee. By the time he was ready to distribute Communion the chapel was a lake, and the noise would have drowned out the world's best choir. You should've seen those beautiful vestments. That rain seemed to turn right angle corners to soak everything, missal, altar linens, everything.

Giving out Communion was slow. Both Padres had trouble. The dampness had softened the Hosts and made them stick together — they had been broken up anyway so as to be sure to go around. And us? Well, Mary, when it came time for me to go up for Communion, I sloshed through water up to my knees. But every last man received. And not one soul left that chapel until the Mass was over, not even the fellows who are not Catholic.

You probably won't like this, but I'm willing to bet not a man in the outfit was homesick just then. After all our work, and the Padre's work, too — the new vestments and all, this Mass ended up in such an unusual way we couldn't think of another thing.

I'll never forget this Christmas. No, not even after we're fifty years married and celebrating with our kids and theirs — I might even tell you this same story every Christmas as long as I live.

Now, you will admit this is a long letter from me — and you remember what I said about exertion in this heat. But I want to tell you the best news of all. There's a rumor of an order that's coming through saying us fellows that's been here the longest can have leaves to go home. If Washington sticks to this line for about ten months more, I figure I'll be due for furlough in time to be with you and the folks next Christmas. Whataya say, baby, do you still wanta wait 'till the war is over, or will you give in and marry me right off IF I get that furlough?

          Yours with lotions of love,

                                  John

Daily Routine on Guadalcanal — for the Padre

One Sunday morning in the spring of 1944 a truly startling piece of news flashed across my vision as I glanced through the *New York Times*. One of the big stores in New York had acquired 500 alarm clocks which they would put on sale next day, "doors open at nine a.m." The following evening the papers implied that the early birds responded in the thousands: police reserves were called out and there were even a few casualties as I recall. I could not help thinking there would be no such commotion over a mere alarm clock on Guadalcanal.

There, according to Father Stack, alarm clocks are non-essentials. Parenthetically, Father Kenneth G. Stack, born in Stockton, California, is the first native son to become a priest, and the first priest to be ordained for the diocese of San Diego, by the first bishop of that diocese, Most Rev. Charles F. Buddy. But here Father Stack's record of *firsts* ceases, for he was not the first Army Chaplain on Guadalcanal, nor the first to write home of his impressions.

On Guadalcanal dawn was generally announced by the thud of bombs falling on Henderson Field; a racket with enough volume to drown out the trifling tinkle of thousands of alarm clocks. Or, if not bombs, then the B-17's, poised on the field for immediate take-off, cleared their morning throats with a deafening roar and whirr as their motors warmed up.

So, the Chaplains on Guadalcanal seldom overslept. After

a quick wash up in the Tenereau River, or elsewhere, they offered Mass, either on the field for the flyers or near some infantrymen preparing for attack. All attest to the fine cooperation of commanding officers in seeing that no Catholic boy need begin a day of danger without Mass and Communion if he desired either.

When Mass was over, Father Stack's next move was toward the end of Henderson Field where, he ironically observes, ". . . ambulance planes unloaded their cargoes of death, meaning ammunition, and reloaded with the wounded to be flown back to safety." Writing for *Extension*, Father Stack complains, "The planes brought in the materials so necessary to kill, and took out the victims of war." At this time of day the sun was barely up but the heat was already intense. Father tried to cheer the wounded awaiting evacuation in the ambulance planes, but the jokes always wilted.

As soon as the planes were in the sky again with the wounded safely aboard, those less fortunate — the dead, required the Chaplain's final attentions. Converging on the cemetery from all directions came trucks of fallen heroes. Each was laid in an individual grave and accorded full burial honors; the Chaplain standing in the midst, blessing each grave and reading the prayers for the dead. This sad task was unfailing daily routine on Guadalcanal. But the Chaplains never conducted it in listless routine fashion. Each burial was surrounded with care, prayer, dignity, and reverence. Each grave was carefully marked, the fallen pilot's wings were attached to the cross bearing his name and the date of his death; the gunner's cross was indicated by a clip of shells. Officers and enlisted men were buried side by side, rank no longer taking precedence.

In the center of this cemetery on Guadalcanal an altar was built of coconut logs. Shell casings served as candlesticks and flower vases. Just beyond the coconut log fence surrounding this American burial ground is that for the Japanese. Here each grave has also been marked with a cross; — "Here lies an unknown Japanese flier," and the date, unless —

In truth this incident occurred on Florida Island, but it may have been duplicated on Guadalcanal:

Navy Chaplain James J. Fitzgerald relates: "One of those hot and sultry afternoons we knew so well there was a call for me to report to the Medical Company. A message like that meant either that a man was very low or he had died that morning. I didn't like the idea of taking a grave marker past the sick bay.

"Upon arriving, I learned I was to bury a Japanese, who had been brought down with two other fliers. One of these had been shot through the wrist, and because he had delayed his surrender, gangrene had set in. From the first, our doctors had held out little hope for his recovery.

"A pitiable sight greeted me at the prisoner tent. The one was dead. Another was reading a Bible printed in Japanese. A working party was called out and a guard of three Marines accompanied us to a near-by plot. A few more Marines gathered around. The working party started to dig the regulation grave. The one Japanese continued to read his Bible. I indicated to the other one that he should mark the name of his dead comrade on the cross.

" 'Engliss?' he queried.

" 'No,' I replied, 'in Japanese.'

"A Navy coxswain dashed down to his Higgins boat and came back with some green paint. With a stick of wood as a brush, the captured flier drew the Japanese characters representing his friend's name. After the grave and the cross were completed I had the working party bring over the body.

"The captured flier seemed to understand for he began to read in a slow monotone. None, except the other Japanese could understand. Yet, with their customary profound reverence in the presence of death, the Marines of the burial party uncovered and bowed their heads.

"The last word was a very distinct 'Amen' in English."

This sad part of the daily routine accomplished, the Chaplain still faced many duties.

One morning young Corporal Joe of the Marines presented himself to the Chaplain who, in this instance, happened to be the beloved Father Frederic P. Gehring — the Chaplain who entertained the men nightly at Admiral Halsey's "Havoc Hall" with his violin. Havoc is the correct word to describe the state of mind of Corporal Joe when he entered Father Gehring's "church."

When he had left home Joe and his young wife had made solemn promises to write every day and love each other forever and ever. Now weeks had passed and Joe had had no word, none that is except a batch of his own faithful letters, all returned marked "Address unknown." When Joe went to Father Gehring, he said not a word, but with grim lips and tear-filled eyes, he released his tight hold upon a batch of his own letters, letting them cascade onto Father's table. The gesture held the threat of disaster similar to that of a falling bomb. Mrs. Joe had moved out on her corporal. Father Gehring sensed instantly that what bombs, bullets, and mosquitoes could not do, the letters had done.

Corporal Joe had taken his squad through tough amphibious skirmishes; he had survived shrapnel wounds and a hard fight with malaria, but it had taken two words, "Address unknown" to break his spirit.

A Catholic boy? No, Joe was a Jew; but he took his load of pain to Father Gehring. It took some time, many words, and much sympathy to mend Joe's broken spirit.

Meanwhile Father Whoever-he-may-be-on-the-post had other details to face in his daily routine on Guadalcanal. The natives called for his ministrations; the Red Cross relied upon him to break the news of home tragedies to the men. With limited means he planned recreational facilities. Father Gehring found time to establish a mimeographed newspaper, *Spaef* (South Pacific A.E.F.). The biggest problem which cropped up unannounced, but often enough to be daily routine, was *nerves*. "They are no respecters of persons," reported Father Gehring, "both officers and men are susceptible."

Men can become hysterical over the loss of comrades; the Chaplain must calm them, even though he too, may have seen a fellow priest fall. Daily he must patiently explain to all comers that he is powerless to speed the arrival of mail; and with every mail call the Chaplain knows that, added to that day's routine, will be a seemingly endless appreciative survey of family snapshots.

Father Gehring's "church" on Guadalcanal was the one spot where the atmosphere and comfort of home could be found. When an air raid or shelling interrupted the daily routine he and his visitors retired to the spacious adjoining fox-hole where Father led them in prayer.

The "church" was built by Navy personnel. A tent covered the framework which was reinforced by a wooden floor and walls. The woodwork was painted a cheerful blue; facing the wide opening was the altar. On one side, running the length of the "church" was Father's living quarters which were partitioned off, naturally. The exit to the fox-hole was here, too. Many a weary lad has rested the night on Father's bunk while he would "be too busy to use it anyway." On the other side of the "church" Father had the walls lined with shelves and filled these with a thousand books. Here, too, were writing desks, always plentifully stocked with paper, pens, and ink — the pens far surpassing the "post office" variety. Other facilities included games, cards, and athletic gear, as well as a phonograph which was in constant use. Often, after a trek to parts unknown, Father had cigarettes for distribution. And because the Japanese considerately installed and then abandoned a refrigeration plant, Father could sometimes dish out ice cream.

On Guadalcanal the Gehring Playhouse prospered, too. Acquiring movie films was just another incident in the daily routine for this tireless Padre. "He has had malaria six times," wrote Barney Ross in a tribute to Father Gehring, "but nothing can hold him down."

In the daily line-up for chow the question "What's new?" usually started the rumor mill grinding, according to Father Stack.

"The hospital is getting nurses."

"Mrs. Roosevelt has arrived."

"Leaves for the whole outfit beginning next week."

"The P.X. will sell beer tonight."

────── so began the daily rumors which the Chaplain was asked to verify. Here, too, the men learned news of their friends: —

"Joe got it last night."

"Yeh, a patrol."

"One of those things."

"The Japs ain't playin' for marbles."

After mid-day chow, next on the Chaplain's schedule was a visit to the field hospital. Here requests came from all sides, the majority of them: —

"Will you write to Mary, Father? Tell her I am okeh. The War Department's telegram will scare the wits out of her."

In passing it might be mentioned that his daily routine seldom gave the Chaplain time for his own correspondence.

After the field hospital visit, another to the ground crews was in order. Here the Padre "sweats it out" with them, waiting the return of the B-17's whose preparations to take-off wakened him in the morning. If the Padre is attached to an infantry or artillery outfit, rather than to the air forces, his daily routine is varied and the tempo, if possible, is increased. He goes where the men go, cheering them on, hearing belated confessions, giving first aid, comforting the dying, helping the evacuation of the wounded — varied, but just another version of the Chaplain's daily round.

If combat circumstances will permit, his schedule includes preparations for afternoon Mass, which is usually offered as near the front lines as possible. On Munda, Father Neil Doyle offered Mass one afternoon in a clearing three hundred yards from the front lines under an arc of artillery exchange. The firing became so intense that the bullets clipped the underbrush close to the altar. Father Doyle finished that Mass on his knees.

One would suppose after months of this daily routine, a pattern not confined exclusively to Guadalcanal, that the Chaplain would welcome a furlough to the States and a transfer to a less strenuous post. But not so. Like the missionary on leave home and surrounded by the comforts of life, "special dishes" for his meals, hot baths, beds with sheets and springs, freedom from the cacophony of mission front or battlefront dangers, the combat Chaplain wants to return. To a man, their refrain is, "I want to get back into action where the war means something. Action is the only thing. When you come back from fighting areas, where every minute means something, this peace and quiet grates on your nerves. I'll be happy when orders come through sending me back on the firing line. *War is my parish.*"

"They Sail Through the Air"

Corporal Jim took a long breath of mild April air. The morning was cool and clear. In another five weeks, when the next batch of volunteers would come out for parachute training, it probably would be hot as blazes. But now, this morning, only the breath of spring was in the air. As he inhaled deeply, Corporal Jim looked over the volunteers gathered in a semi-circle around him.

"So! You fellows want to be paramarines!" he exclaimed, exhaling. "This ain't no strawberry festival," he warned, "it's d—— tough. Have you all taken that little item into consideration?"

The response, as Corporal Jim had expected, was silence. Again he filled his chest with a deep breath and began to walk slowly up and down before the men. He carried the air of a King George or a Churchill inspecting some American lend lease. Presently he paused before one volunteer.

"How much do you weigh?" he asked bluntly.

"One fifty," was the response.

"Um," grunted the Corporal, "there's four thousand men on this station. It ain't possible I can know 'em all, but I'm darned if I don't know your face."

"It's regular issue," replied the volunteer.

There was a twinkle in his eye and a restless movement in the semi-circle as those who knew this volunteer were embarrassed, amused, or both, by Corporal Jim's blustery manner.

"I'll do the wise-cracking," he informed the volunteer.

"I was just thinkin' you don't look like paratroop stuff to me. Ever do any runnin'?"

"Sure," was the quick response.

"Every step you take in this trainin' is on the double," advised Corporal Jim.

"Okeh."

"That goes for all of you," added the Corporal, raising his voice and scanning the faces about him. "Every last move you make is on the double. Don't forget it. Any of you guys caught sauntering along at a fast walking pace climbs the ladder of the flyaway tower six times and trots around this field three times, maybe double three. Get that through your heads. Don't walk — run, an' if you're lookin' for a exit, now's your chance."

None of the volunteers moved.

Then, turning again to the man before him, the Corporal asked, "How old are you, buddy?"

"Thirty-four," came the reply.

"And all in one piece? Holy smoke! When was it you did your runnin'? At Valley Forge?"

"At college — left halfback."

The volunteer might have enlarged a bit. He had always been active in athletics — basketball, football, semi-professional baseball playing catcher and short stop. He had had to earn his way through college, working in the summer at hard manual tasks such as hauling ice, running a lathe, lumberjacking, ditch digging. Parachute training, however, would show his fitness, and in case he could *not* make it, better not go into detail now. To this volunteer it was vitally important that he prove fit.

"Half-back, eh?" he heard Corporal Jim ask. "Look here, buddy, that only takes a little energy. This here is super-superman stuff."

"Are you rejecting me without a trial?" asked the volunteer while another, near the edge of the group tried, by waving his hand, to get Corporal Jim's attention. He succeeded.

"What's ailin' you?" called the Corporal. "It's too early for mosquitoes. I'll be along your way, just hold on." Then to the one he had been questioning, he said, "I ain't rejecting no one without a trial. But still an' all, I know durn tootin' I've seen you somewheres and it wasn't on no football field nor no airplane. You just don't seem to fit here. I'm just workin' on a hunch. But you can start. You'll probably check out tonight."

"Thanks," said the volunteer, grinning.

Corporal Jim started to walk on, paused and came back to his victim.

"I don't get you, buddy. By all that's holy you should be mad as hell by now. I've done my best. But you got a grin on your face. I hold if I can't make a guy mad he's got no guts, an' for this job, buddy, you gotta have 'em."

The volunteer laughed at him, and Corporal Jim walked on puzzled. The men about the victimized volunteer, resentful from the beginning, now laughed too. It was contagious, spreading to the whole group. Corporal Jim flushed, looked about, then stopping before another volunteer, he said, "What are you laughing for? Is this whole batch a bunch of sissies?"

For response the private took the surprised Corporal by the arm, and piloted him back to the man he had been questioning, saying, "You must meet a friend of mine, Corporal. We need a pause for station identification." Then standing before the volunteer the Corporal had questioned, the private said, "Meet Lieutenant Commander . . ."

The Corporal sucked his breath, hesitated an instant eyeing the private to discover if this was some hoax, then he came to attention.

"You've seen the Commander about, right enough," informed the private, "he is the Catholic Chaplain here."

"Why — why — why the h———," stuttered the Corporal, "Why . . . "

"Well, why what?" asked the Chaplain, smiling kindly. "I'm here to learn to jump with the rest of the fellows."

"A Sky Pilot — on the double," said the astonished Corporal.

"And a regular egg," added a Marine standing by.

Rising to the occasion, Corporal Jim said, "I'm sorry, sir, I . . . "

"Forget it," advised the Padre. "You've got a job to do with us. Let's get going. I ask no concessions. If you say I can't make the grade, Corporal, okeh, you're the judge."

The Padre had not volunteered on the spur of the moment. He knew there were hazards in parachuting, but he had learned the paramarines rated a Chaplain and he wanted the job. He wanted to land with them behind enemy lines, carrying the Sacraments, even blessing the very air through which they would descend. If he could take the training he would be the paramarines' Chaplain. Other Chaplains were known as "flying Padres" — why not "jumping Padres," too?

He knew the course had discouraged younger men. Planned to prepare men both physically and mentally for parachuting, he knew it would be extremely strenuous. A sound mind and body were the first essentials. He had both. Could he make the grade? The question was a real one in his own mind — but there he kept it.

To a man, the Leathernecks, themselves, were rooting for the Padre. And the first three weeks were not so bad. The last two would bring the "crisis." At the beginning there were only arduous calisthenics, followed by a run of two or three miles. Then, too, there were lectures, drills, and practice jumping from replicas of planes. Also the volunteers learned how to pack a parachute, and very important, how to tumble: all of this, on the double.

"Hi! Padre!" called a Marine at the conclusion of a hard and tiresome day. Pausing for the Marine to overtake him the Padre debated whether to sit down right there in the roadway. But contemplation of the effort of getting up again deterred him. Every muscle ached, even his eyes ached and his burning feet, it seemed, were trying to burst out of his shoes.

"Thanks for waiting," said the Marine as he came alongside. "How's it going, Padre? I'm ready to hit the hay."

"I'm ready, too," admitted the Padre smiling wearily, "but no can do yet."

"No?"

"There'll be plenty of work waiting for me at the office. We Chaplains have a full schedule."

"You mean you're carrying on your Chaplain duties and doing this stuff, too?"

"That's the size of it," admitted the Padre.

"Well, I'll be . . . "

"Business as usual," smiled the Padre, "daily Mass, confessions, hospital visits, letters home, 'problems,' three Masses on Sunday, daily Office, and extras; the latter in profuse abundance."

"Well, sir," began the astonished Marine, "I'm not a Catholic, sir. I don't go in much for any religion. But I'm here to tell you if I were a Catholic, and you know the Marines are thick with them, I'd go straight to the Pope and see that the Padre had no double duty."

The Chaplain laughed.

"It's my own fault, you know. I needn't have decided to jump with you fellows. But a job is a job. This is a paramarine outfit and I'm the Chaplain. What's the answer? I have to learn the ropes, don't I? I can't be a Chaplain if my men are in the air and I'm anchored on the ground."

"I know what you're driving at," admitted the Marine, "I wonder you get enough out of it to stick with it. I take my hat off to you, sir, and I hope you're aboard every time I make a jump."

Being the senior man of his platoon the Padre was the first man to confront all the dangers of the training. He took the first jump from the 260 foot tower. He was the first, in a harness, head down to be pulled up about 100 feet and then, by a jerk on his rip cord, to be dropped a distance equivalent to the length of his body: this to accustom him to impact.

He was the first to jump from the flyaway tower — a free jump with no guide lines, the student to descend on his own. Just as the Padre's chute was released, the wind changed. Except for his careful manipulations he would have been injured. There were seven more jumps from this tower before the Padre was the first of his platoon to jump from a plane.

Veteran jumpers say the first jump is not so bad; fear, or fear of fear becomes a hazard about the fifth or sixth. The plane was at 1000 feet when the Padre led his platoon in their first jump — and the Marines actually do not jump, they dive.

As all the paramarines in this initial jump made their landing there was a visible and audible feeling of jubilation. They had made the grade, all of them including their Padre.

Over enemy territory, while his men will "dive" loaded with pack and arms and ammunition, the Padre, blessing the air, will lead them, his sole "equipment" being a crucifix, a stole, holy oils of Extreme Unction, and a Mass kit with a small chalice.

PASSED BY CENSOR...

Censored letters from the front often have a staccato quality. Enough news seeps through to be as tantalizing as no news at all. Often, too, it happens that the necessary deletions hopelessly distort the meaning intended by the writer. This is generally due to the latter's failure to absorb fully the instructions on what may and may not be revealed in the written word.

They have been received — those "open and deleted by censor" letters, which may finally reach the addressee something like this ridiculous sample: —

Somewhere in (deleted),
(deleted), 16, 19 (deleted).

Dearest Mom: —

How are you? I am (deleted). We have been here only (deleted) days. The roads getting here were (deleted) with hundreds of (deleted) going back to their homes which we had (deleted). A lota the G.I.'s was hot, tired and (deleted). The scenery here is not so (deleted) but the weather is (deleted). The food is (deleted) but we (deleted) it anyway then our stomicks stopped rumbling the (deleted) code.

Just before being (deleted) to this sector I was to Mass and Communion in a (deleted) hole. That Chaplain was one swell guy. Here we have a new one. I never seen such a (deleted) man. I don't know how he ever has time to sleep, let alone get (deleted) to all the fellows like he does.

[129]

I sure miss you and pop. An' I could do with some of your darnin' on my socks which are practically (deleted). Some buttons on my shirt are '(deleted) in action.' But what the heck? In this (deleted) sector we really don't need no shirts or even (deleted). Only except they do give us some defense against the (deleted) (deleted) what gets at our hides and bite like the very (deleted) himself. (two pages deleted)

Well, Mom, on account of we have orders not to say nothin' of a military nature I must close. Keep your chin up like mine which is (deleted). It is only on account of I know the Censor personally I could write you so much. I am sure he has never (deleted) nothin' I have writ. If he had of I would have (deleted) him.

This place has its cute attractions. If you could send me a couple of bucks I could use them. Over here all the girls have (deleted) eyes and their hair is usually (deleted) too.

Your loving son, Tom.

Such far-fetched distortion is not true of the "passed by censor" excerpts in the following pages. These communications and news flashes have been more carefully written and so give a wide variety of news from all fronts. If our imaginary Tom should come across these, he may profitably observe what "military information" he may safely send "Mom."

* * *

WAS THIS YOUR BLOOD?

Overseas, 1944

"Dear Friends:—

"Some days ago you will have received notification from the War Department of Pat's death; but that official telegram will not have given you the full details, which, as you read them now, will surely ease the pain of this parting and fill your hearts with gratitude for the mercy of God and the ceaseless efforts of the American Red Cross to secure blood donations for our brave boys.

"This letter comes to you from Pat's chaplain, but in reality is truly a personal message to you, his parents, from across the abyss that separates this world from the Eternal Realm of Almighty God.

"Everyone liked Pat — a starry-eyed lad to whom all was adventure. As you know he had just turned twenty-one. He was not a bad boy, but like every son of Adam there were a few things he had to 'straighten out' before he should be fully ready to meet his Creator and Judge.

"It was 01:00 hours. The earth lay bathed in a pool of silver. Suddenly the sharp crack of a rifle split the still air and Pat fell to the ground seriously wounded.

"He was rushed to the hospital where, for three long hours, he hovered unconscious between life and death. Though doctors and nurses used their every skill to save him, it was apparent to all that they were waging a losing battle. In the busy hush of that operating room I began the prayers for the dying.

"Moments passed. In a final desperate effort the doctor called for some plasma and prepared to give your dying son a transfusion. As the blood secured by the American Red Cross — perhaps, friends, in your very own home town — perhaps your own blood it was — in any event as this blood coursed through Pat's veins his pulse grew stronger, and presently with a fluttering of his eyelids Pat awoke to full and clear use of his faculties for twenty precious minutes.

"In that time, my friends, he made a full Confession and received Our Lord in Holy Communion and the Last Rites of the Church. Then, at peace with God, he lapsed into unconsciousness and died.

"True the blood plasma did not save the physical life of your beloved son, but what is infinitely greater to Pat and to you his sorrowing parents, it made secure beyond any doubt his eternal life with God.

"May these details bring you comfort and consolation and may the Peace of God fill your hearts.

"Sincerely in Christ,
"Joseph J. Walsh (Capt. ChC. USA)
"(Diocese of Pueblo)"

* * *

BREAD FROM HEAVEN....

"Somewhere in England, 1944.

"Dear Mom: —

"Your last letter received, also the package. Thanks for each. The snaps of sis are swell. She is getting out of her pigtails I see, and I suppose you have given in now and let her have a lipstick. Why not?

"One thing I did not like about your letter, Mom. Quit worrying about the jumping. Yesterday I made my twenty-seventh jump and I am writing you now, see? I'm keeping straight with God as you say — so what is the worry? You know this paratroop outfit has its own chaplain, Father John S. Maloney from Blessed Sacrament parish in Rochester, New York — he jumps with us. What more do you ask, Mom?

"And here's a line for you. The chaplains and the doctors attached to paratroop units do not have to jump with us — some regulation excuses them, but Father volunteered to serve with paratroops and was with us in training in the States and he insisted he be allowed to jump. He has a very adventurous spirit, Mom, and I can tell you the fellows think jumping is okeh when the padre bails out with the rest of us. And here's the payoff. It will hand you a laugh. It seems when Father Maloney was in the seminary a doctor told him he had a weak heart and ordered him not to play baseball and not to take any strenuous exercise.

"And here's another slant for you, too. When the big moment comes, and we jump somewhere else besides somewhere in England like we do now, the padre will be along. We will be weighted down with more junk than you have stowed away in the attic, but not the padre. But he will have What Counts! He will carry the Blessed Sacrament. Down from the skies will come us, Mom, and floating along with us our Chaplain and Christ, Our Lord, Himself. 'Member about 'the Living Bread comes down from heaven'? We'll be coming with It. Now, Mom, get wise and lay off your worrying. What can happen to us?

"Please send some decent tasting toothpaste in your next package.

"Your loving son,
Sam."

* * *

GERMANY IS A BEAUTIFUL COUNTRY . . .

To Chaplain Stephen W. Kane, (Captain, U.S.A.), of Des Moines, Iowa: —

"For gallantry in action. Chaplain Kane received a report that the body of a member of his regiment was lying in a mine field. The body had not been previously recovered due to the danger of encountering an unexploded mine. Captain Kane, with complete disregard for his own welfare, voluntarily went into the mine field and recovered the body. The courage, bravery, and devotion to duty displayed by Chaplain Kane are deserving of the highest praise and reflect the finest tradition of the armed forces.

"By command of Major-General Ward."

The *Award of the Silver Star*, September 9, 1943.

"Oflag 64, Germany,
November 17, 1943.

"This strange address has its own story . . . finally the Germans got me in the close of the Tunis show, so here I am, sweating out the days in a German prison camp. I am most fortunate in my share of luck since many of my boys were asked to make the supreme sacrifice. . . . Life itself may not be so intensely valuable, but somehow we do love life!

". . . I have not had any U. S. mail since Christmas, so I am very much in the dark, but then we shall see you all before long for a little while enroute for the Japanese show.

"There is not much that one can write about on this side. Prison life is dreary — one of petty tyrannies, but we see the skies and bless our stars. Germany is a beautiful country, and barbed wire cannot spoil the entire grandeur of a land in war. Europe is altogether such a change from the desert lands in Africa.

"Ours is a boundless, unshakable confidence that justice and peace will soon bless our world.

"Father Steve Kane."

* * *

NUNS TAKE IN WASHING . . .

Chaplain Edward J. Waters (Major, U. S. A.) of Rochester, writes from "Someplace in Sicily, September 9, 1943. . . .

"Yesterday was a day of great rejoicing for us all. Italy signed an Armistice, and I received my first mail in weeks.

"The natives appeared most happy at the word of Italy's surrender. Crowds gathered in the streets in a festive mood and all the church bells told the glad news for these unfortunate people.

"Our troops took it all in stride. No one became alarmed as they looked toward the future and knew that much more blood and sweat must be given before the day of final victory.

* * *

"The unfortunate civilians in Germany and Italy have my pity. They usually suffer more intensely than our troops. To live in constant terror of the bomber will produce many nervous people after this war.

"One Sunday morning I passed through a small Sicilian city immediately following a German raid. Those who were not dead or wounded were hysterical. The Devil invented the bomber.

"My troops and I made a gift of $537 last Sunday to the nuns who live in the Benedictine monastery about five miles from us. The nuns were forced to invest this money in Italian bonds — so, still they are penniless.

"Due to what is here called a 'traffic jam' I was stopped near their chapel, and I heard familiar music and the words *Tantum Ergo*. I entered and remained for Benediction. After this I stopped at the grille to speak to the nuns: I had a premonition that they could use a friend. They told me they were in great distress. So . . .

"I arranged for several officers to send their laundry to the monastery. Now the nuns earn about $100 per week in the 'laundry business' and the business increases each week. Yesterday the Superior told me that if we stay in these parts just one more week, she will have ample funds to maintain the monastery for one year.

"The nuns are praying daily for the troops and for me."

* * *

"... TERRIBLE EVILS, BUT ..."

In Sicily,
3 September, 1943

"The Most Reverend Karl J. Alter, D.D.
Bishop of Toledo,
Toledo, Ohio, U. S. A.
Most Reverend dear Bishop:

"Autumn is almost upon us. The Bishop is aware of all the events in North Africa and of the more recent invasion of the Sicilian isle. . . .

"Like Africa, Sicily is now history though the island was not without its moments. On July 10th I went over the side with some of our infantry. In those immediate hours which followed the initial attack my Mass kit was left at the beach head. It was not until the western portion of the island was taken that it finally reached me. The fact is, however, during that period Mass was an impossibility.

"Since the fall of Messina certain privileges have been granted the troops. One day on a tour of historic places we visited the cathedral of Monreale. Here is rebuttal to the narrow gibbery, the ignorance of the scoffers of the 'dark ages.' It is a prayer in itself; magnificent — it has soul. The remains of several of the early Norman kings are here. In the chapel of the Blessed Sacrament is a simple marble casement sealing the remains of the crusader Louis IX of France who died of the plague in Tunis — hundreds of years before the thunder of Allied and Axis might crashed in that same theater.

"Your Excellency should know of my late good fortune. At our headquarters there is a chapel (even this is several hundred years old) completely furnished and equipped. This is an experience — so radical a change it seems more a dream than a reality. His Eminence, the Cardinal Archbishop in Sicily has granted Father McPartland and myself all faculties. Every morning to Mass comes soldier and peasant, Count and beggar, young and old. And about our soldiers — this war has occasioned terrible evils, but men who in their Faith before were diffident have emerged from these experiences with a knowledge of what is Catholic Action — and they are so minded. Here is the solid rock foundation which bids fair for the Church in America in the years ahead. I know of no priest in the service in this theater who has not marveled in admiration at the conduct and example of these men; and himself not been inspired therefrom.

"Through the past ten months all of us ever more and more have been made to realize and appreciate God's loving Providence and Our Lady's intercession: we remain constant in prayer. Again all of us here commend ourselves to your prayers and to those of our people. . . .

"Sincerely in Xto
Urban J. Wurm
Hq., 2nd Armored Division
APO, New York"

* * *

"... JUST AN AVERAGE CATHOLIC BOY ..."

Somewhere in Wales

"Dear Aunt Mae: —

"You were asking about Mass and confessions in your last letter. Well, I can say that prior to Army life I was just an average Catholic boy but since I've been away from my family and my wife for so long I've been a little better than that. I guess we all turn to God in our necessities. But the Chaplains make it so easy for us to be good Catholics — Confessions at any time, Masses every day, one in the morning and one in the evening. And best of all we must only fast from liquids one hour and solids three hours before going to Communion.

"Four of us got together here a few weeks back, and on our own time, we built a rather nice altar and altar steps and a Communion rail for our chapel. We presented it to Father (William J., S.J.) Heavey in time for New Year's Mass. Just looking at his face when he saw what we had done was enough gratitude for me. He is such a swell guy we would do anything for him. Over half of our outfit is Catholic. A sergeant from Chicago and the 1st sergeant from Pittsburgh, and myself are making a drive for attendance at an initial Holy Name Society Communion. So far so good.

"Your nephew,
William Burke."

* * *

"... MY FIRST ELECTRIC LIGHT ..."

(As written to Father Edward Dougery by Father Joseph McGoldrick, both Chaplains, formerly of Los Angeles.)

"Personally I am feeling fine and dandy. I am doing my best to convert the heathens but that is rather soft, in that there are no marriages or baptisms. Masses are offered at odd hours and fasting at times takes quite a beating; but being in a combat zone you never can tell when a 500-pounder will land in your hip pocket.

"I cannot tell you how long we have been in our present position, but it has been long enough. Anyway we have not had a hot bath in many months. I saw my first electric light in a long time the other day and it sure did look funny. We can say 'we have contacted the enemy.' For further information see your local newspaper.

"Since arrival I have been underground at least one third

of the time. Getting to feel like a groundhog. To make things pleasant, try this: some black, wet night, jump out of bed through the mosquito net and head for the fox hole. It won't be where you expect it, but there will be a puddle of water. Try again and after hours it seems, you have it with about eight inches of the wettest, stickiest and blackest mud known.

"Or maybe you would rather try this: Someday when you want a smoke open a pack, there they are, not wet, just moist with a touch of the prettiest green on the end. They call it mold at home. Cut off the mold and get the matches — just damp enough so they won't light; then either find someone with a light or throw the cigarette away.

"We have developed quite a bit of proficiency in washing our laundry and our faces in a tin helmet. It is quite an art to get a pair of overalls in a helmet and scrub them."

* * *

A LITTLE BOMB — A LOT OF RELIGION

(From Chaplain M. M. Tennessen, lieutenant of the Navy Chaplain corps, to his Bishop, the Most Reverend Gerald T. Bergan of Des Moines, Iowa.)

"A daily duty of the chaplain is his visits to the (hospital) wards. On weekdays this duty is usually performed from 09:00 to 11:00, and likewise in the afternoon 14:00 to 16:00. There is no telling the many tasks that may come his way in this duty. Sometimes the writing of letters for men who cannot write, telling the folks that 'all is well and a letter will follow.' The distributions of reading material, papers, magazines, tooth-brushes, razors and blades, and when he can do so the delightful handing out of candy and chocolate bars. His most reliable source of supply is the American Red Cross. Prayerbooks and New Testaments are always in demand. Rehabilitating these men is oftentimes a difficult task.

"But nowhere in any hospital back here is the performance of these duties accompanied with a greater measure of gratitude. There is no moaning or groaning, regardless of the degree of suffering. Most patients are all smiles. And the most deserving are the very ones who ask for nothing. Grateful for a glass of water and glorying in the luxury of a mattress and sheets as well as clean, spotless pajamas.

"Sometimes it happens that word is received in advance of new arrivals. The loud speaker sounds the call of 'all hands to quarters.' Then the chaplain's place is at the receiving ward, in readiness to administer to the casualties and also to help the doctors and hospital corpsmen in the work of salvaging life from the havoc of war. This work will continue all day and all night, until every patient has had everything that modern surgery and medicine can do for him. Plasma administration is working wonders. A call over the loud speaker for donors of blood is immediately responded to by the hospital corpsmen, all rushing to the laboratory to give their best.

"Movies are the greatest enjoyment of all the armed forces. They are the regular feature in most of the camps in this area of the South Pacific. But of greater importance to the morale of our men is the mail call bringing letters from home. Almost unbelievable is the excellent performance of Uncle Sam in getting the mail to all these far-flung corners of this global war. The mail call cannot be daily, but it is surely regular. And when it is irregular, there is a good reason for it.

"Going through the ward one day I had a copy of the Memphis, Tenn., paper. In the first ward, I asked the first patient where his home was. 'Memphis, Tenn.' was the answer. As I peeled off the top copy of the paper from his own home town he almost developed shellshock. Strange things happen. Yesterday I noticed a paper in an office adjoining mine from my own home town, Kenosha, Wisconsin. I inquired around and found another shipmate from Kenosha who had been receiving this paper regularly.

"Back in the States, every church and probably every pastor was witness to the loss of faith and leakage from the fold, especially in the growing generation of our youth. Not so in so-called war zones. A little bomb and its reverberation, sending all into fox-holes or dugouts, carries a religious call instead of the church bells or chimes. And a bomb will do the impossible especially if it sounds on a Saturday night. The next day many a new face will be seen at divine services, a face that will tell the story of a sleepless night: bombs increase church attendance. Men are learning to live new lives out here, — and it is so easy with all the opportunities that the Army and Navy offer.

"Our men are finding out it is just as easy to bless as it is to damn; that it is better to pray than to curse, and that a clear conscience is good equipment to carry when the 'zero hour' strikes or an 'alert' is sirened.

"A chaplain's life is a great adventure. He can accomplish more good in the South Pacific in one month's time, than he may ever have dreamed of in parish life back home. For there is many a boy who depended upon his mother to say his prayers, but out here he is learning to say his own."

* * *

WHITE VESTMENTS — JUNGLE MUD . . .

"A suggestion for you, Mother," wrote Chaplain Arthur K. Duaire, "in case your electric washing machine breaks down — but do not set up this operation anywhere but in the wide open spaces!

"Jungle mud is hard on white vestments and the cleaning process very long. I wanted a quicker method. First I talked the mess sergeant out of an empty five gallon food can. Next I begged wire from the communications sergeant. Then, punch a hole in either end of the can, attach the wire and suspend between two trees — plentiful on the South Pacific island where I devised this experiment. Suspend the can so that its bottom is a foot from the ground. Next fill a small can three fourths full of sandy gravel and saturate with a used diesel oil (not rationed). Heat the oil to combustion point and ignite by placing burning paper atop the oil. Fill the top can with soap powder and water, douse the clothes and smoke cigarettes for three hours, then the washing's done. This is an excellent process in the jungle, but it could be improved with blueing and starch."

Father Douaire, an Army Captain, formerly of Chicago, was cited for bravery under fire in landings in the southwest Pacific from October 27 to November, 1943. "Chaplain Douaire was instrumental in the safe evacuation of the wounded and administered the last sacraments to the dying while he was exposed constantly to enemy fire. He arranged and conducted burial services for the dead."

* * *

"ONE G.I. STAYED KNEELING . . ."

(A portion of a letter to his wife, written by a soldier "somewhere in England.")

"Several weeks ago a Mission was being held by Father Gallagher (Chaplain John Gallagher, C.Ss.R.), at his station. Several hundred boys were in a big barracks on the edge of a field. Father Gallagher was saying Mass and all the boys were going to receive Communion.

"The alert came on and an assisting priest told all the

boys to run to shelters. Sam and others had to dive into a culvert and were soaked. Two big ones shook the ground. They come in threes.

"The 'all-clear' came. More than ever thankful to God, the boys returned to the blacked out hall. There, with two altar candles burning, was Father Gallagher who had remained at the altar, oblivious to all. He turned with the Host and every boy cried at his courage and devotion. One G.I. had stayed kneeling — a tough Sgt. altar boy.

"The sequel — the next day, the third bomb was found, 50 feet away — a DUD. Was this a miracle!

"Father Gallagher only smiled when I told him the wonderful story about himself, but he praised the altar boy who had refused to leave. The G.I. Sgt. said to him, 'Father, how could one die better than serving God like this?'

"The building was almost lifted off the ground. The 500 lb. DUD would have obliterated it."

Chaplain Matthew Meighan, C.SS.R., writing of this incident says the "lad" brought the souvenir to him — but still it *was* a dud!

THE PROTESTANT BOYS MADE BETS . . .

"With a tank for my backdrop I offered Mass out near the front lines in a driving downpour. . . . It seems the Protestant boys of the unit had made bets with the Catholics that I'd never show up in such a torrent. My lads collected the bets. . . .

". . . We came to a spot in the thick of the jungle. . . . Knowing that with these fool Seabees, a trail this morning can grow into a six-lane highway by nightfall, I picked out a site for my chapel . . . there was a little matter of jungle vines, undergrowth and tall mahogany trees to be removed. But to a Construction Battalion that was like mowing the front lawn. Early the next morning an operator drove his bulldozer smack into the heart of the jungle, and presto! my chapel site was a clearing. Then came the erection of the chapel — this time a hospital tent . . . and the building of an altar . . . speedily the painters whipped off a sign for me on discarded boxes of K ration:

HOLY MASS

Daily, 6 P.M.
Sunday 8 A.M., 4:30 and 6 P.M.
Chaplain Father James Rice."

"MADE TO UNPACK"

"A few of our soldiers here in England built a confessional out of cast-off packing cases. The padre is delighted with it, especially for the remains of a painted notice on a board visible on the penitent's side. It reads: 'MADE TO UNPACK.' I've left it there as a polite hint to come clean!

Chaplain John Powers"

FUNERAL IN DUNGAREES

A transport was torpedoed off French Morocco; Chaplain Francis J. Ballinger (USN) escaped, on a raft. He had just finished Mass as the blow came. Immediately his attention was given to the wounded and to assisting them into life boats; then carrying the Blessed Sacrament, he climbed down a cargo net and into the sea. He lost all his personal possessions and his vestments, and later had to assist at a funeral attired in dungarees!

"THY WILL BE DONE"

The crash of his plane landed him in a Burma jungle (Chaplain Harry F. Wade, C.SS.R. Army Captain). The plane lost its way over the high Himalayas and all hands were ordered to parachute. Father went crashing through some trees on his descent, but escaped without injury. "After six days in the jungle in uncomfortable proximity with wild elephants and tigers, I was rescued. I kept repeating over and over, 'Thy will be done' — and strangely enough I was unafraid."

THE FOURTH BATTALION

Technical Sgt. Samuel Shaffer, U. S. Marine Combat Correspondent, has a high regard for our Chaplains. "Father O'Neill was worth as much to the boys of his Marine regiment as a trainload of ammunition," is the way he puts it, writing about Chaplain William R. O'Neill of New York and a Navy Chaplain. Sgt. Shaffer was with Father O'Neill for sixteen months almost wholly spent in the Pacific war zone.

Sgt. Shaffer, a Jew, reports: "There is no greater single contributing factor to the very high morale of the troops than the spiritual guidance the boys are getting from the chaplains . . . there is not a man in my regiment who would not gladly lay down his life for Father O'Neill, Jew, Catholic or Protestant. He talks the language of the boys; he lives their life and is ready any time to help any boy whether the trouble be spiritual or worldly.

"Father organized a group of men in the regiment and called them 'the Fourth Battalion' — the 'miracle' of that Battalion will long be talked of. It was also known as 'Our Lady's Battalion' and was made up of men who promised to say their Rosaries and wear medallions of Our Lady. All were very devout and faithful to their pledge. The battalion was organized just before we moved in to Guadalcanal, and when the battle was over not a single Marine of 'Our Lady's Battalion' was wounded."

Reminding the reader that one of the jobs of the Chaplain is to write letters to parents and next-of-kin of those who have been killed in action, Sgt. Shaffer continues, "Imagine the job Father O'Neill had after the Battle of Tarawa. I can tell you he never wrote two letters alike. There was no form letter stuff for him. To each one he wrote straight from the heart. . . . And another thing: when the Jewish Holy Days approached, Father made arrangements to have the Jewish boys of the outfit given liberty so they could attend services at a place nearby, where he had arranged with a Rabbi to care for them.

"The most moving sight I ever saw was on Christmas Eve, 1942. We were at a South Pacific port waiting sailing orders . . . there was no liberty for anyone. I don't know where he got them but Father O'Neill showed up with an organ, a victrola and some records of Christmas hymns and carols. He said Midnight Mass in a dingy old shed at the wharf and it was about the most beautiful thing I ever saw."

As a native of Washington, D. C., and former reporter for the *Washington Times-Herald*, Sgt. Shaffer had made it a practice for a number of years to attend Midnight Mass at the National Shrine of the Immaculate Conception: he could appreciate the contrast of a "dingy shed" on the water front.

FORTY-FIVE CONFIRMED

From "somewhere in New Guinea" Sgt. Robert M. Hopkins of Gloversville, N. Y., wrote his mother: "Tonight at 7:00 p.m. we had a very beautiful ceremony. It was the dedication of our new chapel, Our Lady of Victory . . . tonight's ceremonies also marked the first time that the

Sacrament of Confirmation was conferred by a Catholic Army Chaplain (Fr. Gearhard of the 5th Air Force) in this area of the Southwest Pacific. He had a special delegation from the Apostolic Delegate to Australia.

"There were about 45 confirmed, including both officers and enlisted men. All in all there must have been seven or eight hundred Catholics present and I think I was one of the most proud. Eight of the fellows confirmed received their instructions from me. I was also sponsor for two of them.

"For the rest of my life I shall always remember tonight. It was more than touching. It really epitomized the real reason for all this struggle in the world today — the fight to make it livable for a Christian way of life. Right is right and wrong is wrong. Hitler and his satellites would like us to believe wrong is right . . . but truth is one and unchangeable."

"WE EXPECTED TO BE ANNIHILATED . . ."

"I went into Tarawa with the first battalion of the second regiment," writes Father Joseph E. Wieber of Niles, Michigan, to his parents in Lansing. "The battle of Tarawa was fought on the little island of Betio in an area about three quarters of a square mile. . . . Only two chaplains reached the island on the first day. Very few of the men who left the ship in our Higgins boat escaped death or injury. In fact, very few live and uninjured Marines were on the island that night.

"I started toward the beach in the command boat of our assault wave. We had to transfer to an amphibious tractor under heavy shell fire . . . so heavy that we took the command flag down. I was the only officer on the amphibious tractor and so was in command.

"The tractor was put out of commission by a shell and we

considered wading in: but we decided it was too far and waited for another tractor. That little wait probably saved our lives. After we landed we loaded the tractor with wounded and sent it back to the transport. This gives an idea of the confusion that prevailed and the demand for individual initiative.

"The first night on the island was a critical one. We expected to be annihilated and would have been had the Japs been on their toes and counter-attacked. The following day, reinforcements which could not get in the day before finally landed and things looked better. The fighting during the first two days was terrific. The first three days of my seven day stay on the island were spent giving the wounded spiritual and physical aid. So many corpsmen and stretcher bearers were killed that we hardly got a breathing spell during all that time. The fact that we were constantly under rifle and machine gun fire made it even more difficult. Men were killed right in our 'sick bay.'

"It was impossible to give the Last Sacraments to all the wounded and dying. But the Catholic Chaplains, both Navy and Marine, gave all the Catholics an opportunity for confession and Communion while they were on board ship before the battle. . . . This last experience has made even more evident the importance of having everything under control spiritually before leaving the ship for a landing.

"If God had not given me His special protection I would have been wounded or killed at least a dozen times."

FOUR HUNDRED YANKS . . . ON THE SANDS

Even after two years of war Americans do not fully appreciate the inestimable value of blood plasma, nor the necessity of building up vast reserves of it. In World War I sixty per cent of the injured died. Thanks to blood plasma plus sulfa drugs only one per cent of our men wounded on Guadalcanal died. In the early days of the North African invasion 400 critically wounded Americans were practically given up as lost. When plasma was administered, all but eight of the 400 survived. Some special guardian angel must have been guarding those 400 Yanks bleeding on the sands of Africa, because the plasma that saved them was *flown in from America at the last minute.*

(What's your score, fellow American? Are you a member of the one or two gallon club?)

"... IT'S PERFECTLY SAFE"

Father Joseph E. Whelan, formerly an assistant pastor of a Kalamazoo, Michigan, parish after emphatically denying the rumor that he was killed in a plane crash, told a story of a burying detail. The place was in the jungle where he states, "It was a common sight to see our men with their rosary about their open necks, reading their missals or New Testaments all along the front — in slit trenches or behind trees — waiting for an attack. Many times after a bombing or strafing attack the men would grin and say to me: 'that's the best Act of Contrition I've said in a long time.'"

But about the burying detail . . .

"The commanding officer instructed me to tell the detail that the area to which we were going to bury some men was free from enemy fire. I obeyed. I did better. I told them that in effect our field of action would be just as safe as if we were back home in bed.

"Off we went, but soon we were compelled to hug mother earth. After our arrival at the burying spot we alternated digging graves with digging ourselves in a grave at five or ten minute intervals. We did this to show our respect for the 'over' fire of the enemy which he was directing our way.

Finally the sergeant of our detail picked up a piece of hot shrapnel which plunked in the fresh earth between us. He tossed it from hand to hand, and grinned at me: 'Oh, sure, fellows, it's perfectly safe, nothing to worry about at all — as safe as if you were home in bed!' We all laughed and went back to work. The men exercised as much care with the last grave as they did with the first.

"The commanding officer had been right, of course, but an area may be safe one moment and not so safe a day later."

PERHAPS YOU KNOW...

REV. GERALD WHELAN, C.SS.R., who was torpedoed in the North Atlantic on the same occasion when REV. JAMES M. LISTON lost his life; or another Redemptorist, REV. JOHN G. SCHULTZ, C.SS.R., or REV. JOSEPH HRDLICKA, C.SS.R., who wrote home: "Our only consolation is our tremendous work for God on twenty-four hour shifts in soiled and sweaty clothes...." Running short of altar breads he baked his own "... knee-deep in mud, with rains swooshing over me, with insects gobbling at my hide...."

Still with the Redemptorists, perhaps you know REV. ROBERT HEARN, C.SS.R., formerly a missionary in Brazil, who in far off Australia met an old friend of seminary days CHAPLAIN ARTHUR FINAN, C.SS.R., of the U. S. Navy. "Business as usual in a foxhole," writes Father Hearn.

Perhaps you know Mrs. Sullivan: "Sure, we're Catholic, and Irish, too! Five wounds, sure, five heart-scalding wounds; but didn't Christ suffer five wounds? And His Mother standing right there to watch Him die. God's Will be done. They were good boys, all five of them...."

When FATHER JOHN S. WISE, C.SS.R., signed up he chose the Army. "The Navy has its points," he admitted, "but the ocean is too deep for swimming." When REV. DANIEL O'BRIEN, C.SS.R., reported to his Commanding officer on American territory in the Atlantic within fifteen minutes he was being flown to San Juan to conduct a funeral.

Perhaps you know one or more of these Army Chaplains: — REV. ANTHONY G. VAN BEERSUM of the Diocese of Nashville; REV. GEORGE J. FLANIGEN, former editor of the

Tennessee Register, who has seen duty in Australia, REV. WALTER O'BRIEN, O.F.M.Cap., of Brooklyn. The Diocese of Springfield, Illinois, has given eight priests as chaplains: the REVS. JOSEPH DINEEN, HENRY SCHWENER (Navy), ALOYSIUS SCHWARTZ, L. J. MCDONALD, PHILIP J. NEWMAN, JOHN B. DAY, CASIMIR ANDRUSKEVITCH and JAMES J. HAGGERTY. REV. TERRENCE BRADY was killed in action in the Pacific Theater.

Perhaps you know that FATHER DE SMET was asked to act as a chaplain with the U. S. Army in 1858. He left St. Louis University to comply with the President's request and was assigned to the army of Utah. — That over 150 priests from the Archdiocese of Chicago enlisted as chaplains before the second anniversary of Pearl Harbor. — That REV. GEORGE M. KEMPKER is serving with the U. S. Marines, that REV. THOMAS M. REARDON (U.S.N.) was the first priest to land with the Marines on Guadalcanal — he was featured in a book and a motion picture, was the victim of malaria and blackwater fever and invalided home where he was assigned to duty at the Naval Base Hospital, Brooklyn.

Did you know in the Army our Chaplains hold the ranks from Lieutenant to Colonel in deference to the cloth? That the Chief of Chaplains, MONSIGNOR ARNOLD, is so by presidential appointment, and holds the rank of Brigadier General by special Act of Congress? That the first afternoon Mass was offered June 15, 1942, at 6 p.m. at the U. S. Marine Post, Quantico, Virginia, CHAPLAIN PAUL J. REDMOND, O.P., celebrating? That thirty received Communion after a four-hour fast? And did you know the FATHER KEMPKER mentioned received the Silver Star Award for bravery — on Bougainville? The battle of Piva Forks was his third on Bougainville. Perhaps you know REV. OZIAS B. COOK of the Archdiocese of Los Angeles who wrote a bereaved mother of the death of her flier son. Father Cook buried him, and covered the grave with orchids. Or REV. THOMAS O'MALLEY of the same Archdiocese, who wrote home: "Pray like you have never prayed before . . . for the men going into battle, for the men in battle, for the wounded, the dying, those going back into the lines after a short rest." REV. JUSTIN E. FREEMAN, O.S.B., thinks the Military Ordinariate should have medals for the Catholic boys who win other souls of other nations to the True Faith, as Sgt. Jim Boland won a Chinese boy named Gee. REV. FRANCIS CAHILL, a naval chaplain, from Peoria, Ill., says, "I've lost all my respect for Illinois wind . . . at one of my bases the anemometer (a wind gauge) broke after it reached 135 knots." The letter was written from a base in the Aleutians. From the same diocese REV. MICHAEL HADDIGAN, formerly superintendent of schools in the Peoria Diocese writes from the South Pacific, that he found a naval base fifty percent Catholic with no Catholic Chaplain; he chose to visit a leper colony for the good of his soul; now he has haunting memories of those unfortunates.

Studying the works of the martyred priests and sisters of St. Mary's mission in Ruavatu, CHAPLAIN JOHN W. SCANNELL of the Army mastered sufficient Guadalcanalese to solemnize the nuptials of Solomon Island couples. On March 28, 1943, he gained the title of "Marryin' Padre of Guadalcanal"; on April 16 of the same year he was awarded the Silver Star — not for his record of marriages, but rather for his bravery in burying the dead under fire and cheering the living. When REV. FRANCIS X. MURPHY of the diocese of Little Rock returned home on furlough he had a collection of souvenirs representing the handicraft of the natives of Woodlark Island: the natives were mostly Catholic. . . . "Hey, Jim, it's all right, there is a chaplain here . . ." he was REV. JULIUS S. BUSSE serving on Attu who heard the men on the left flank had no chaplain and went over to see what he could do for them. While he was helping the wounded; two bullets passed through his raincoat and another smashed his eyeglasses. Perhaps you know the Navy chaplain Rev. John Burns from Davenport, Iowa, or Rev. Vincent Walsh who served at Treasure Island among the remains of the World's Fair, and carried on successful convert work, or Rev. Thomas J. Wolfe, also of Davenport? Or perhaps you know Chaplain Richard Scully, alumnus of St. Bernard's Seminary, who relieved monotony on a Pacific Island by organizing a softball league; or, Rev. Francis J. Keenan, C.M., who won two awards for bravery, the Silver Star and the Purple Heart; or Rev. Francis Gorman formerly of St. Elizabeth's Hospital, Chicago. Now the padre dodges Japanese bullets on Bougainville . . . officers are prize targets for the enemy.

Perhaps you know that Lt.-Cmdr. Chaplain John P. Murphy had to show himself around his old haunts — the diocese of Lincoln, Nebraska — to disprove the story of his death. Or Rev. B. V. Schomer of Des Moines, Iowa, who saw nine of his Marine converts confirmed in the "Pacific Theater"; or Rev. James Gilloegly, attached to the U. S. 14th Air Force in China. A Maryknoll missioner of Scranton, Father Gilloegly baptized a Chinese infant brought into the world by an emergency road side operation. Rev. Thomas I. Conerty of Brooklyn is much loved by his Marines in and about Cape Gloucester, New Britain. Another in those ports is Rev. Francis X. Shannon whose boys built a chapel in the jungles and dedicated the altar in memory of the mother of Archbishop Spellman of New York. "There is nothing vague about the religious attitude of America's fighting men," says Rev. Francis V. Sullivan, "It is a solid, definite thing." Chaplain Sylvester Wagner, orginally from Chicago, was lauded in the post paper as, "The Fighting Chaplain" — of the Navy. He contracted a lung condition as the result of his service with the C.B.'s in the Aleutians. Chaplain C. J. Yeager serves the new Rainbow

Unit, and Rev. Thomas Tooher of the Albany Diocese is highly popular according to the *March Field Beacon* which says he has served "with justice and humanity." Rev. William M. Slavin takes care of C.B.'s; and the senior Catholic Chaplain in England is the Rev. John E. Foley, a native of Memphis.

Perhaps you know that the consensus among our Army and Navy Chaplains is that 38% of the Army is Catholic, at least 50% of the Navy, and more than 50% of the Marine Corps — seems as though Catholics take to water, from Baptism on. Perhaps you know that 25,000 Catholics fought for Independence in 1776 and more than a million shouldered arms for Uncle Sam in World War I. You've probably read about Chaplain Stephen J. Meany, S.J., who has told home front audiences our boys are fighting to get home. His medal of the Blessed Virgin was struck by a sniper's bullet on Makin and ricocheted through his chest and arm. Perhaps you know Rev. Richard D. Power of San Francisco whose men collected $2,000 to pay off a church debt in North Africa. Perhaps you know the casualities, Rev. John Joseph McGarrity, who lost his life in the battle of the Java Sea, and Rev. Thomas J. Knox, of Savannah who died in this country.

Did you know Rev. Charles H. Parks of New York was the first Naval chaplain? That was back in 1888. Or Rev. James J. Fitzgerald of the Navy, who was commended for his service with the Marines at Tulagi? Or Rev. Matthew F. Keough, one of two Navy chaplains privileged to wear the Presidential Unit Citation ribbon awarded to the First Marine Division; and he also has a letter of commendation and the Purple Heart Medal. "Long Term" Navy chaplains are, Revs. Thomas F. Regan of Milwaukee, William A. Maguire of Newark, and John J. Brady of New York.

Perhaps you know Rev. Stanley J. Kusman, S.M., serving in Italy. In ten days and nights under enemy fire he buried American and German dead. A Catholic officer of New Zealand wrote the parents of CHAPLAIN MENSTER of Dubuque " . . . your son has spared no effort . . . a Catholic priest is truly a father to all the boys from all the countries." Or, do you know REV. JOHN L. CALLAHAN of St. Louis, who was killed in a glider crash in North Carolina? You must know REV. JOHN LOUIS BONN, now a Naval chaplain, formerly of the faculty of Boston College and interested in the Catholic Theater Movement as well as author of the fiction best sellers: *So Falls the Elm Tree* and *And Down the Days*.

Perhaps you know REV. JAMES J. O'DONNELL of San Francisco serving in New Guinea after his initial months at Monterey Reception Center in California. One of his men from Connecticut says he is "one of the best liked officers here . . . popular with all faiths. . . . " Or do you know REV. AMBROSE SULLIVAN of Indianapolis, one of the forty-two Father Sullivans in the services. The Sullivans are second in number to the Murphys. He was for a time at Port of Spain, Trinidad, after spending a year on Aruba, a Dutch island off the coast of Venezuela. The novelty of listening to Negroes speak with an English accent is a major diversion for his men which does not wear off.

Or perhaps you know REV. FRANCIS J. MCMANUS (Lt. U. S. N.) of Cleveland, chaplain aboard the USS *Canopus* when she was bombed in Mariveles Harbor, Bataan, . . . also a Silver Star chaplain and a prisoner of the Japanese. Or, REV. JOHN L. CURRAN, O.P., of the Savannah-Atlanta Diocese and REV. JOSEPH V. LAFLEUR of Lafayette Diocese, both recipients of the Distinguished Service Cross and also prisoners of the Japanese; or, REV. JAMES W. O'BRIEN of San Francisco who wrote from his Philippine prison camp that he enjoyed excellent health.

You may know REV. EDWARD J. WATERS of the Rochester

Diocese who was cited by Brigadier Clift Andrus and Major General Terry Allen who state: " . . . he was present with batteries under attack by air, artillery and bombs, administering to the wounded and dead . . . in addition to his normal duties he . . . (assisted) in burying enemy dead." This was in the African campaign. Or, perhaps you know, REV. JAMES S. MCGINNIS, S.J., of Cleveland who had to persistently urge his superiors that he was not too old to volunteer. He was cited for bravery on Guadalcanal. Or, REV. MICHAEL J. LYONS of the Syracuse Diocese. For service in the Far East he received the Legion of Merit Award; so, too, did REV. JOHN J. WOOD, C.S.V., of Philadelphia in the Southwest Pacific. Citations for exceptional devotion to duty have been received by REV. AUBREY ZELLENS, O.S.B., of St. John's Abbey, Collegeville, Minnesota. Do you know the four-times-decorated chaplain, REV. ALBERT STEFFENS? Or, REV. ANSELM M. KEEFE, O.Praem., De Pere, Wisconsin, or REV. FRANCIS W. KELLY, Marine Corps Chaplain, and REV. GERVASE SHERWOOD of Los Angeles? These three have received Legion of Merit Awards.

Perhaps you know REV. S. ERNEST WILEY, first priest of the Diocese of Nashville to enter the Navy. Or, from the same diocese, you may know one or all of the following: REVS. CHARLES QUEST, FRANCIS REILLY, and ROBERT WILEY. Perhaps you know REV. ANTHONY A. ZIOBER formerly of Chicago, and since "somewhere in Italy" where he received a promotion in rank; or, two other Chicagoans, now Army Captains, REVS. WILLIAM J. BENNETT and JOHN J. WALL — or yet another, REV. GERALD M. DOUGHERTY, O.S.M., who has been spiritual adviser for Catholic nurses in Africa. Still in the windy city — do you know, REV. HAROLD P. O'GARA? He enlisted the day after Pearl Harbor, is on foreign service and has written a history of the 149th Infantry in World War I, the Mexican and Civil Wars. He is making history with the 149th in World War II.

Hopping over to Davenport, Iowa, perhaps you know some of these chaplains: —

Rev. J. E. Toomey, Captain, 91st Infantry Division. At Camp White a native said to him: "The less said about this part of Oregon, the better." — or

Rev. Herman Strub, Captain, 1st Evacuation Hospital — or

Rev. F. J. Deschenes, also with an Army Hospital — or

Rev. Kenneth C. Martin, 316th Station Hospital (Davenport seems to have a specialty in serving hospitals!) — for,

Rev. Harry B. Crimmins is also with an Army Hospital in North Africa.

Other Chaplains from Davenport Diocese are: Revs. Clifford A. Egert, William T. O'Connor, E. H. Ruhl, Edward L. Lew, and Rev. Martin J. Diamond with the 117th Station Hospital who met another from his own town, Rev. F. J. Lollich in Australia. With the 201st Infantry is Msgr. J. D. Conway. Or perhaps you know the Revs. Patrick J. Toomey, C.S.V., or Richard J. Eagan.

Some more Navy chaplains whom you may know are: Revs. John P. Wissert, C.PP.S., and Henry Barge, C.PP.S. Rev. E. A. Moorman of the same Order is with the Army — so, too, is the secular Rev. Ralph A. Thompson. Rev. Frank J. Barry is retired from the Army with the rank of Lieutenant Colonel. He was with our troops in Northern Ireland when illness forced him home. He began the trip on the SS *Wakefield*, but she caught fire at sea and Father Barry had to make a quick transfer to another ship in mid-ocean.

Did you know Rev. Edwin Ronan, C.P., is a prisoner of the Japanese; that Rev. John McAuliffe is on Pacific duty; that Rev. Austin Henry of the Army is back in the States after serving in the Pacific area for more than two years; that Rev. Clement A. Siwinski of the Diocese of

Lincoln was in India when he was advised of his promotion to Lieutenant Colonel? Officially his address is Washington, D. C., but as supervising Chaplain of the Army Air Force Command mail may reach him in almost any spot over the globe. REV. LAWRENCE F. OBRIST of the Army was first at Amchitka, then later removed to Anchorage, Alaska. Fellow alumni of St. Francis Seminary, Milwaukee, REV. JOSEPH NIGLIS (Milwaukee) and REV. FRANK GEIGEL (Green Bay) were in the Aleutians at the same time as Father Obrist. The latter hails from Lincoln, Neb., as does REV. JOHN P. MURPHY who was wounded in active combat duty in the South Pacific.

Here are chaplain prisoners on Bataan, not elsewhere mentioned: perhaps you know REVS. HERMAN C. BAUMANN, Pittsburgh, ALBERT BRAUN, O.F.M., RICHARD E. CARBERRY, Portland, Oregon; WILLIAM T. CUMMINGS, M.M.; JOHN E. DUFFY, Toledo; JOHN J. DUGAN, S.J.; HUGH F. KENNEDY, S.J.; JOHN J. MCDONNELL, Brooklyn; EUGENE J. O'KEEFE, S.J.; STANLEY J. REILLY, San Francisco; THOMAS SCECINA, Indianapolis; HENRY B. STOBER, Covington, Ky.; ALBERT D. TALBOT, S.S.; JOSEPH VANDERHEIDEN, O.S.B.; MATTHIAS E. ZERFAS, Milwaukee.

Perhaps you know REV. FRANCIS EARLY of the diocese of Indianapolis, an Army Chaplain, who won commendation for his recovery of the bodies of men killed in a plane crash in North Africa. But he is far more interested in a chapel he acquired for his men "somewhere in Egypt."

The commanding officer of his regiment allowed him to re-locate a barracks for his chapel and add any new construction necessary "without trying to make a cathedral."

"Every bit, building as well as all interior fixtures," writes Father Early, "were made by the soldiers. They put their hearts in the work. . . . The altar is made of wood but it fools everyone with its marble-like finish. The communion rail is finished like the altar; the lamps are made of ply-

wood and clear glass frosted by the Army painters. ....
We have started the novena in honor of the Miraculous
Medal and at the first service we had 250 men. I ran out
of books and medals.

"Chaplain Rixey of the chief of chaplains office told me
this was the nicest chapel he had seen under Army jurisdiction."

In a "word for parents" Father Gehring wrote of the
Rev. James Dunford who gathers the boys around his
tent each night for the rosary — "another Father Duffy at
Guadalcanal," adding, "To the mothers and fathers of our
boys in the service I would say, kneel down in prayer and
thank God for chaplains like Father Jim." If not personally,
then by reputation you must know Rev. Bernard Hubbard,
S.J., who up Alaska way solved the problem of warmth
without weight for the Air Force. He invented a quilted
eiderdown suit, copied from a Mongolian Eskimo garment,
which reduced the weight of our aviator's flying suits from
fifty to *seven and a half pounds*.

Did you know that in North Africa, General Mark W.
Clark presented 25,000 lire which had been donated by his
troops to Msgr. Francesco Guazzo, Vicar Forane of Vallo-
Capaccio? The money was to be used to repair a tower of
the ancient church of Maria del Granato which was built
in the tenth century and damaged by the Americans as they
routed the Germans from the town.

After the enemy was routed it was Chaplain Patrick J.
Ryan of the Archdiocese of St. Paul who suggested our men
repair the damage to the tower of the venerable church.

General Clark has an interesting keepsake of the occasion.
Msgr. Guazzo presented him with a carnelian on which
is carved the image of Neptune. The temple of Neptune,
erected by the ancient Greeks, stands on a plain within
sight of the damaged church.

Did you know four hundred alumni of "Boys Town"

donned the uniform? That three of these, out of the first hundred to see combat duty, won the Silver Star Medal, and a fourth the D.S.C. posthumously? That ten, of the first hundred, were killed in action? Perhaps you know Rev. Cyril Feisst, now in India where an Army Major serves his daily Mass, another Chaplain who has written appreciative letters to his home diocese, Spokane, of the work of the Catholic Missions. Perhaps you know Rev. Anthony J. Conway of Philadelphia, a Navy Chaplain, or Rev. Denis G. Moore of the Archdiocese of San Francisco, and Rev. George T. Quinlivan of Albany, both serving in Italy and points north. Do you know Rev. Stanley C. Brach, Army Chaplain, a prisoner in Germany? Or Rev. John V. Loughlin of the Rochester Diocese, Navy Chaplain, cited for bravery by Admiral Nimitz after the Tarawa fight? Or, from the same diocese Rev. Elmer Heindel, praised by the Vicar Apostolic of the South Solomons. "He is still the priest you know, full of zeal, high spirit and good sense of humor," so wrote Bishop Jean Marie Aubin to the Most Rev. James E. Kearney, D.D.

"Men against cold," is the warfare Rev. Wilfred Bouchey of the Albany diocese knows from his post in Greenland. Father Bouchey had the distinction of offering the first High Mass in Greenland in more than 500 years. Rev. Leo LeSage of Ottawa, with the Canadian forces in Sicily, utilized the hot tropical sun to dry his laundry as he shaved! "Four minutes for a pair of pajamas," he reported.

Can it be, you have read thus far and have yet to find a priestly name familiar to you? Don't give up. Follow through.

Try some of these: — perhaps you know Rev. John T. Fournie, who reported at the Army Chaplains' School early in the spring. He is only one of several from the Diocese of Belleville — such as, Rev. Harry P. Mannion, and

Rev. Joseph Duehren who is with the Airborne Troops in England; Rev. J. Hugh Kilfoil also in England, Rev. Thomas P. Driscoll of the Marines who went along in the Marshalls invasion. Or, from the same diocese, do you know Rev. Charles A. Nebel, or Rev. Vincent Tikuisis? The senior chaplain from the Belleville diocese, Rev. Julius Babst, died in October, 1943. Did you know the first priest to hit Namur in the Marshalls invasion was Rev. Emmett T. Michaels of the diocese of Altoona? When he stepped on the beach a land mine exploded twenty feet away, but he was uninjured. Also "in" on the Marshalls invasion were the Revs. Thomas V. Brody of Chicago, John M. Dupuis, C.S.C., formerly at Notre Dame, and Roderick Hurley, O.Carm., formerly stationed in Chicago.

Perhaps you know Rev. Richard Scully, formerly of New Britain, Conn., whom the boys called the "No. 1 man on the island" — in the Pacific; or Rev. John N. Landry, also from Connecticut. Or do you know Rev. Joseph P. Owens of Union City, N. J., who turned the sympathy of a Panama Canal community from pro-Axis to pro-United Nations "by simply performing the duties of a Catholic priest!" Or do you know another early summer "graduate" from Harvard, Rev. William Van Garsse of the Diocese of San Diego? His ability to speak six languages has been valuable to the Army. The Californians find the North Atlantic climate a little "tough" according to Rev. Daniel J. Ryan of San Diego.

Perhaps you know this hit-or-miss, pick-and-choose chapter must come to a conclusion. Perhaps you echo the "miss" and "choose." Of course it could go on, and on, and on, like the creek to the brook to the river to the sea. Here again apologies are offered for omissions, and to show good faith in spite of these, the following paragraphs, written by Rev. Lawrence R. Schmieder, U. S. Navy Chaplain, are offered for contemplation:

"The missionary and the Chaplain in combat zones make wonderful meditations. They have nothing but eternity in prospect for them and that perhaps on the double. It is surprising how much better one can meditate when one realizes that the end of our mortal life may be very near. Going through one or more of these experiences leaves an indelible imprint on the mind and heart. We realize (in this very personal way) that we have 'here no lasting city but seek one that is above.' I have always admired a missionary and a military Chaplain going out for his second 'hitch' in the 'combat zones.' They realize very vividly what the thoughts and feelings of Our Lord were as he set out for Jerusalem the last time — for His Passion and Death.

"People who have never lived through actual modern combat just do not know what it is all about. The procedure can be described to them, the movies can simulate scenes and sounds, but the effects on their minds and hearts are not the same. I have always admired the pioneer, the man on the frontier, the man on the fighting front. People on the homefront never underestimate their own importance, but in a war, the final decision is made at the fighting fronts."

PERTINENT MISCELLANY

BY NOW the reader will appreciate that there was no exaggeration in my "Thank You" chapter at the beginning of this work. If anything I am guilty of understatement there. For here I am, almost at the limit of my allotted number of pages, yet still my work table groans under a pile of pertinent miscellany. What to discard, what to include has been the problem all through to this point. Now I am reluctant to conclude without presenting all these interesting items: — yet even so this work still will remain a remembrance of omissions! Shall the following long bits and short bits be also sacrificed to space limitations? I know now that the task of a compiler has its drawbacks.

First among the items I would include here are excerpts from a letter written to his friends back home by Army Chaplain Frederick G. Dorn from "somewhere in Oahu." Previous letters had spoken of his hope to visit Molokai. At last this desire was realized. A plane from Oahu to Molokai was "a little jump over the Pacific blue made in a short time."

"From the airfield we drove by motor to the top of the great divide which separates the little peninsula Kalaupapa from the other part of Molokai Island. This 1400 ft. mountain wall fences off the leper colony settlement on one side, the Pacific ocean bounds it on all other sides. No vehicles go down the leper side of the mountain wall, not even a bicycle. The descent is made on foot or by horseback or mule pack. The trail is narrow, steep, rocky and winding along the high cliffs of the Pali in serpentine fashion.

"Below at the hospital Dr. Sloan took my companion and myself in tow and with him we toured the settlement which is under the territorial government and financed by the same. At present there are 350 patients and approximately 75 'clean' personnel — staff members, employees, Sisters, Brothers and so forth. It is a well regulated community with all modern conveniences. Required supplies come to the inhabitants by sampan. The patient-citizens have their own laundries, bakeries, machine, paint and carpenter shops, store, beer and light wine shops, two jails, courthouse, judge, sheriff, four policemen, a theater with two shows a week, recreation centers, post office, schools, a fine Catholic Church with a resident French priest, Father Peter d'Orgeval, S.S.C.C. — a very spry man of 72 — a Protestant Church with a resident minister and a Mormon meeting place with someone coming in from the outside to conduct services.

"Some forty-odd serious cases are in the hospital; others less advanced are in their own dormitories with their common dining rooms. Still others live in private homes — some of them are married. If the latter have children, these are taken away at birth, although they need not necessarily be afflicted with the malady. Some married partners never contract the disease. Patients who are in a position, or condition, to work can earn regular wages. The sums expended by these people for war bonds and Red Cross drives went into astonishing figures.

"With Father Peter I visited the Bishop Home for Girls conducted by the Franciscan Sisters (Third Order, Syracuse, N. Y.). They have a nice little chapel of their own. At the hospital I saw various stages of the disease. You have read descriptions and seen pictures of the deformities. I saw some almost normal patients walking about, others working or sitting around — others were in bed in rather extreme condition. Very deformed and disfigured faces and limbs left a deep impression on me. Scenes of sorely afflicted people I shall not soon forget. They were listening to the radio, smoking, reading or playing games. One bed patient was playing chess with another who sat by the bedside. The bandaged stumps of the fingers of the first held a pincers with which he managed to move the pawns.

"Another memory of my visit is a charming and cheerful old lady. She is 73, has been on Molokai 57 years, even recalling the last six months of Father Damien's life among the lepers. I found her sitting on the floor near her bed. One leg is off at the knee, another has a deformed foot. She is also blind. Yet this white-haired, crippled blind old lady seemed joy personified. She is a convert: every evidence of her conduct, good humor, contentment and real happiness seemed to testify to her spiritual insight.

"At the other end of the settlement Father Peter and I visited the Baldwin Home, a Government Institution, where the Brothers of the Congregation of the Sacred Heart help the male population. Here we found agriculture. There was a herd of beef cattle some distance away. Chickens are fed coconuts, split in half. They thrive on this fare.

"I shall not soon lose the impressions of that memorable day with the lepers — the friends of the saintly Father Damien and of his Master, the Saviour of Lepers."

Next in collection of pertinent miscellany to be included here, is a letter describing a tour of our Pacific bases. It, too, records impressions, though of a vastly different nature than those of Father Dorn. Frank M. Folsom, Vice-President in charge of RCA Victor Division, is the author of the following paragraphs. They record his impressions acquired during a period when he was a Special Assistant to the Under-Secretary of the Navy and Chief of Procurement. The trip here described was made in September, 1942, in the party of the now Secretary of the Navy, James V. Forrestal.

"The Itinerary took us from Washington to San Francisco, to Pearl Harbor via clipper, to Midway and return to Pearl Harbor via Marine Transport plane, to New Hebrides, the Solomons, and New Caledonia via Navy four-motored patrol bombers, and after inspecting the islands in this section we returned, via the Fiji Islands, the Samoa group and a few intermediate stops to Pearl Harbor and finally home.

"While at Pearl Harbor I was the guest of Rear Admiral Robert English, Commander-in-Chief of the Submarine Fleet in the Pacific, who met an untimely death in a routine transportation accident. We arrived in Pearl Harbor on a Friday and were there over Sunday. On inquiring where Mass might be heard, I was told it could be heard in several places, but the nearest was the Submarine Base, and this is where I attended Mass on Sunday morning.

"Mass on the Submarine Base was held in the movie theater, and it seemed very strange to kneel on the floor and assist at Mass in this slanting position. A lasting impression of the occasion was the chaplain, sitting on a common kitchen chair in the middle of the stage, hearing confessions before Mass, with his arm on the back of his chair, his head in his hand and his eyes closed. The men stood in line, waiting their turn: there was no screen, no formal confessional, just the priest sitting on a chair on the stage. Officers and men stood together in line and when they came to where he sat, knelt down and made their confession in front of a full movie theater.

"When the chaplain finished hearing confessions, a portable altar was rolled on the stage by two bluejackets, who properly arranged it, lit the candles and then served the Mass. The theater was full, the enlisted personnel mostly in blue denim work clothes, the nurses in hospital uniform, officers in work and dress blues — all serious and thoughtful. In the group were some men who had just come in on a sub after having been out for seventy-two days. They all looked well and I learned that many non-Catholics in the group were also giving thanks for their safe return after the hazardous trip which they had just completed in a very dangerous combat zone.

"On Midway Island, the Commanding Officer was a Navy Captain, "Beauty" Martin. He received the nickname at Annapolis, and it has always stuck with him. He was a fine officer and his command comprised the Army and the Marines as well as the Navy. One of the striking evidences of solidarity in the combat zone is that often all three services are under a common command, sometimes Army, sometimes Navy, sometimes Marine.

"Captain Martin gave the Under-Secretary and our group a reception in the Officers' Club which was an old Pan-American Airways waiting room. The men had covered the walls with drawings of gooney birds, ferns and other local inhabitants so that the place had a club atmosphere.

"In the group gathered for the reception, I noticed one of the officers wore a cross on his collar which of course would not necessarily indicate whether he was a 'Ch.C' (Catholic Chaplain) or a 'C' (Chaplain) which might be any of the Protestant denominations. But I went up to him and said, 'I judge you are a priest.' He replied that he was; that his name was Father Francis Wollack,* and that he was from the diocese of Rochester. He had been out there several months and had lived through the famous battle of Midway.

"I asked him some questions about this engagement and he told me on that occasion he had not had his clothes off for more than fifty hours. In every case, the men who were going out, many not to return, asked him to come around and give them his blessing: often leaving with him keepsakes to send home to some loved one. He told me with tears in his eyes of the fine faith of all of the men, regardless of their formal religion, and of their devotion to their coun-

---

* EDITOR'S NOTE: This was undoubtedly Rev. John F. Woloch.

try — their willingness to die for it, which many of them did.

"There are two islands in that group, and Father Wollack offered two Masses on each island on Sunday, and one on each island on weekdays.

"He was the confidant of both officers and men, the liaison officer between different groups, the arbitrator of disputes of all kinds, and he won his way into everyone's heart through his faith and devotion to duty. I have written to him several times and know that he has been transferred to another island in the South Pacific but, of course, he cannot say just where. He has been in that general area for the past two years.

"There is another chaplain who is outstanding, but about him I can say little because of the nature of the duty to which he has been assigned. He was on a sub-tender in the mid-Pacific — the friend of some nine hundred men, one hundred and fifty of whom were Catholics. His name is Father Slattery* and as I remember he was from Trenton. He is a young, thin, fine looking Irishman with a very winning smile. The Skipper of this particular craft told me that everybody on the ship had a great affection for Father Slattery; that he was the type of chap who was a born leader, and all liked him instinctively.

"At one of the island stopping places between Pearl Harbor and Sydney, I met Father Matthews, a Major in the Army. At this tropical place they never have any rain and there are no trees or shade but lots of sun and lots of heat twenty-four hours a day. They sleep with a sheet for covering and water was rationed by the simple expedient of turning it off at the source; all water, of course, having to be distilled out of sea water and put into tanks for use by the personnel.

"Father Matthews was a man about fifty-five, and he

---
* EDITOR'S NOTE: Is this Father Edward A. Slattery of the Newark Archdiocese?

[170]

looked as though he might have been the pastor of a small mid-western parish. He wore shorts, no tie, sleeves rolled up, but somehow you could tell by the kindly gleam in his eye that he was a priest. I remember walking up to him and saying 'Good morning, Father.' He looked at me in surprise and asked how I knew he was a priest. I said, 'Father, I'm psychic; I just couldn't miss.' He seemed very pleased that I had been so sure of what he was without needing any preliminary questions.

"Father Matthews was doing a great job on this base. The men were here temporarily on their way toward the battle front. All were anxious and expectant to get going. If they stayed very long undoubtedly they would have become irritable, impatient and difficult to handle. Father Matthews knew them all, he had their confidence and he was a splendid influence on that tiny tropical island. He, too, was anxious to go forward to take his share of the front line burden. 'I don't want to sit back and let the other fellows do the whole job,' he told me. And in those words he epitomized the spirit of all the chaplains whom I met on this tour of the Pacific."

So Mr. Folsom discovered first-hand, so to say, what this whole volume has tried to explain, namely that the Catholic Chaplain is doing all he can to make the Sacraments and the aids of our religion available to the Catholic laity in uniform. The Church has, through the Holy See, given the Chaplain many privileges, too, which aid in his task.

Many and varied, as has been shown, are the experiences of our Chaplains. One day the service in a post guardhouse was finished and the Chaplain turned to the men, suggesting, "If any of you are Catholic you might want to talk to me a few minutes. I'll be in the office for awhile."

The first-comer was a tall, well built boy with an outdoor look in his eyes. "I'm here, Father," he began uneasily, "for being picked up AWOL. Really I was runnin' away."

Father nodded. The thing to get at, he said to himself, is to find what this lad was running away to. So he asked, "What part of the country are you from?"

"Minnesota," said the boy.

His eyes had a far-away look. "You woudn't know the town," he told the chaplain. "It's a little old place — on a lake."

"Farmer's son, eh?" asked the Chaplain.

"Yes, Father. My folks own the best farm round about."

"And what is going on on the farm now?"

"Thrashin'. And dad's got a new thrashin' machine." The boy snapped his eyes from the far-away vision of his home, to fasten them on the Chaplain. "With me gone," he said seriously, "there ain't nobody to run that thrashin' machine."

"I see," said Father, smiling at the Catholic soldier. "You figured you'd make a run for Minnesota and the thrashing machine. Even if it meant AWOL, eh?"

"That's it, Father," said the boy, honestly, "I didn't want to desert."

It had not occurred to this lad to arrange a furlough for "thrashin' time" — but it is one of the Chaplain's duties to find out what goes on inside a soldier.

Possibly separation is the prime heartache of war. With all it is not intensified by "thrashin' time" — but there is a time every year when the heartache on all fronts, including the home front, increases painfully — that is Christmastime.

Elsewhere we have had an account of Christmas Mass in a Pacific jungle. Here follows an account from Father William J. Hayes, O.F.M., of midnight Mass offered for his men who then were on manuevers "somewhere in Tennessee."

"Midnight Mass was the greatest thrill I have experienced. In the pouring rain men came walking, wading through the fields as early as ten o'clock. I started a big bonfire at 10:30 and those men stood around in the rain and mud to go to Confession. I sat in a jeep covered with blankets, my stole over my neck, and heard confessions from 10:45 to 12:30. The Mass did not start until I finished and when I went over to get vested in the rain there were over three hundred men waiting. Six boys held up blankets over the jeep where the altar was set and they held a blanket over me, too. Then I began the Holy Sacrifice, offered to God in the rain and mud and sleet far in the hills of Tennessee.

"I had to preach despite the rain — and moved over to the fire for that as I was very cold. The soft, homesick voices of three hundred men singing 'Silent Night' so touched me I could hardly go on with the Mass. Inspiring is the word to fit this occasion. How pleased Our Lord must have been to see this sight: over one hundred and fifty of those men knelt in the mud to receive Holy Communion."

In the European theater this same example of faith and love has been enacted on many scattered fronts. It was on a Christmas Eve that Chaplain James C. Lane managed to sing his first high Mass in two years. It was in the vicinity of Tunisia. "It had poured all day but finally cleared up about 6 p.m. and was not too cold. This was a break for us as the Mass was to be offered out of doors.

"A bugler called Church call exactly at midnight and the Mass began. The music was wonderful. I used the P.A. system to preach and my voice kept re-echoing off the surrounding hills. My friends, the Italian priests, attended and helped me distribute Holy Communion. The place was packed."

On another sector of the Italian front, where Chaplain John D. St. John, S.J., discovered what it is to sleep in mud and wake up surrounded by water, midnight Mass was offered by him in a beautiful church. "It had been miraculously spared from destruction when, sometime ago while

Mass was being offered, a bomb crashed through the roof but did not explode.

"I was kept busy for hours before Mass hearing confessions. During Mass the man sang feelingly and stayed afterwards to continue singing Christmas carols.

"For my second Mass I went in my jeep to one of my Fighter Groups some miles away and offered the Holy Sacrifice in a cold, dark and damp stable. The only light came from candles and a flashlight focused on the Missal by a G.I. It was a most impressive sight and one I shall not forget soon. The men had just returned from a sweep over the lines and were still in their heavy flying clothes. It hit me hard when they came up to receive the Prince of Peace in their hearts. Just back from killing although there isn't a murderer amongst them, and now receiving the Source of all Life. The whole atmosphere made us feel closer to the Babe of Bethlehem because of the conditions closely resembling the first Christmas — a stable, cold, damp, dark and the same Babe lying on the Corporal.

"The day ended up with evening Mass at a Heavy Bombardment Group forty miles away. I got back to my post wet, muddy and tired, but happy, and I wouldn't exchange places with any Chaplain in a Post Chapel in the States."

Realizing the fullness of the Chaplain's life, one wonders sometimes how he finds time to write home such complete accounts of his activities. But, with the rank and file a little time at least is generally available for recreation and letter writing. So it happens, that just here, I am able to include excerpts from two servicemen whose impressions merit recording and who have written at greater length than most of our Chaplains.

Yeoman William J. O'Grady found himself in company with four thousand naval personnel in a North African city on Christmas Eve. Writing his parents, he told them:

"Over 4500 bluejackets jammed the North African cathedral. It was a Solemn High Mass offered by an English-speaking priest of the Cathedral with Navy Chaplains acting as deacon and sub-deacon. The Bishop attended as well as some high officials, civilian and military. By a special dispensation of the Bishop, Father Paschal E. Kerwin, O.F.M., of Boston, granted Absolution to all the servicemen from the pulpit. This absolving grant — the first time I ever heard of it — is given by chaplains for large numbers of men in battle areas. During the absolution all bow their heads and recite the Act of Contrition, privately. Then Father Kerwin gave a short sermon.

"Throughout it was a wonderful sight, but the most joyous spectacle beyond doubt was the Communion, when row upon row of Navy men, neatly attired in their blue-dress uniforms, rose in a body as it were, to approach the rail to receive the Body of Christ. The Bishop had high praise for all the men, with special emphasis on the thousands of communicants. The collection at that midnight Mass amounted to $7,000 which was more money than had been collected in that cathedral for a whole year. That figure just about floored the Bishop."

The following morning, according to his letter, William O'Grady attended a second Christmas Mass in a convent chapel. "The celebrant was Father Kerwin. At the Introit prayers I noticed many nuns answering the responses aloud. It confused me for a minute, but they soon lowered their voices to private prayer. Possibly they believed I wasn't thorough enough in serving, but they discovered differently. The Mass was said in the presence of the Spanish Council. . . . After the Mass the children's choir sang Christmas Carols accompanied by the clickety-clack of castanets.

"Once again I say I had a wonderful day, — but deep down in all our hearts there was something missing — our homes and our people."

Another serviceman, Tom Houghland of the Army,

bluntly admitted in the very beginning of his Christmas letter home that he was homesick and forlorn as Christmas approached and he found himself on foreign soil in enemy territory. But he found time on Christmas Day, 1942, somewhere in North Africa to write,

"This is my first Christmas away from home. Like the rest of the Yanks I am lonely and homesick — terribly so — but there has been a midnight Mass, and that helped. The sky here in Africa was its natural shade, black velvet studded with winking stars and the night was cold, too. But we were all glad it was not a white Christmas because none of us wanted it to be like Christmas at home.

"Mass was held in one of the hangars, the best repaired one. It was not unlike a cathedral with its shadowy corners and arched ceiling far above us. In one corner stood the crib built by the willing hands of a certain line crew. The little figures were there, standing in the straw, the same as they are today at home in St. Ambrose. More than this, above the crib there was a gaping hole in the wall torn by a bomb during the campaign. Through the hole I could see the black Arabian night — the same night with the same stars that stood over the original Crib in Bethlehem almost 2000 years ago. I thought of all these things as we knelt there in front of the stable, and somehow it seemed to me we weren't quite so far away from home.

"Trimmed with branches of evergreen trees whose ancestry probably goes back to seeds blown by a dusty wind from the cedars of Lebanon to this country, the altar stood nearly at the rear of the hangar, bathed with flood lights which were landing lights removed from old planes. It was covered with snowy linen borrowed from the French chapel not far away. The French chaplain was the celebrant, the English chaplain the deacon and our own chaplain the subdeacon. An RAF choir sang the Mass of the Angels, and I think it was perhaps the best singing I have ever heard.

"At the Offertory we all sang *Adestes Fideles* and I felt a great lump in my throat, hearing all those strong voices about me raised in what is probably one of the favorite hymns of everyone. As I looked about me I saw that nearly everybody's eyes were glistening with tears and I knew that we were all thinking of the same thing — home, and the last time we'd heard that. There were hundreds who went to Communion and knelt on the dusty, oily floor to offer their thanksgiving.

"As I walked back afterward I thought to myself that was one *of the nicest Christmases, or rather holiest Christmases I have ever* known, and I shall remember it always. Stripped of the more personal things we have known such as mistletoe, holly and tinsel, Christmas was more real to me than ever before. Everyone felt that way. I heard one boy say, 'I don't think I've ever felt closer to God than I did during that Mass.'"

When we read an expression from the heart, such as the foregoing, who may deny war holds some blessings for those who will pluck them. It is hard for those of us safely at home, kneeling in our parish churches, to visualize Mass in rain and mud, in foxhole and jungle. The very thought spells distraction, — and if we are addicted now to such — what can it be like to hear Mass under truly distracting circumstances?

Father Alphonsus J. Paulekas, C.SS.R., mentioned a variety of distractions in a letter describing his "super de luxe open-air chapel" in New Guinea.

"The entire village came to Mass. This statement also includes the babies. I'm sure I said a valid Mass, but must admit there were a few distractions. Babies were crying, mothers shushing, youngsters restless, a man leading them all in prayer in a strange language, dogs running around and picking fights with each other, and mosquitoes taking frequent and substantial nips."

The jungle holds no monopoly on distractions, however. One Sunday morning in sunny Italy, according to the *Chaplain's Digest* published by the University of Notre Dame, Father Joseph Barry in white vestments was saying Mass in an open field.

"Father Joe turned around for a *Dominus Vobiscum*, saw his congregation scattering in all directions, and immediately heard the Luftwaffe overhead. Father Barry made a lightning review of his moral theology, decided that to interrupt the Sacrifice at that point was permitted, and took off for a bit of cover himself. The nearest foxhole was a good distance away, the F-W's were coming on wide open, and — plunk! — into a mud puddle went amice, alb, cincture, maniple, stole, chasuble and chaplain. The Mass was resumed later in dappled vestments."

When the Catholic laity under arms continues to "come back for more" in the face of such discomforts and distractions, it would hardly be fair to boldly state their Catholicity has been strengthened by the war and its circumstances. Surely only a strong Catholicity would persevere under such varied but generally trying conditions. War may have increased the devotion of many, but the parents who wait the return of their sons and daughters deserve credit for rearing them in a practice of their faith. Even on the battlefront the lax and the fallen-away are present in proportionate numbers, but the preponderance of evidence, even the opinions of some few Chaplains to the contrary notwithstanding, seems to indicate a virile faith among the majority. Naturally a Chaplain's major concern is to bring back the fallen-away.

Perhaps it was the fallen-away whom Father Bernard Gannon of the Diocese of Los Angeles and stationed at the Army Air Base at Pyote, Texas, had in mind when he inserted these thoughts in a bulletin which he published for the cadets.

"What will the priest think when he hears the load I have to tell? He will think: — How unworthy I am of such a privilege as this! To be able to send this man out walking on air, forever free of the load he has been carrying! . . . What a time this lad must have had getting up the courage to come to confession! . . . I wonder if I would have had the guts to go to confession if I were in his shoes? . . . I'll bet it's his mother's prayers that obtained the grace for him . . . she's probably been making novenas for this for months. . . . For the chance to hear this one confession I would wait a lifetime. Thanks be to God."

"His mother's prayers" . . . much of the Faith in the armed forces — the strong, virile Catholicism rests on "his mother's prayers." There probably is not a mother alive who would not step in front of an enemy bullet to protect her son. Nor a father. But mothers and fathers cannot be out there where their sons are exposed to death — physical death.

But there *is* a Mother who can be there. God's mother. And the way to reach her, to ask her to protect these boys, is to simply ask. A wounded marine, writing his mother noted that his watch had stopped at 1:45, the time of the explosion which knocked him out. "Nobody but us will believe it," he wrote, "but that is just a quarter to eight back home, when you were receiving Communion for me at St. Cyril's. I know now that God is going to have the two of us together again. Keep up the Holy Communions."

That is the assistance our Chaplains ask of the folks back home. Prayer and Holy Communion. They will tell us, it is true, that an Army marches on its stomach; we know this too, else why have we had rationing at home? Our Chaplains beg us not to extend the rationing to our prayers. We may be thankful to our country — when other countries, some of them, display such anti-God "samples" — that she has realized the military must have more than beef and butter, ships, tanks, planes and guns to make her invincible. In the long run, it may be God's own design, that our Armed Forces are only as strong as their faith in Him. And so: — *Oremus.* God in His mercy, in His good time, and for His great glory, will do the rest.

* * *

THE CHAPLAIN'S PRAYER

Dear Lord, King of Peace, I, your priest who was ordained to bring peace to men's hearts, find myself in the armed service of my beloved country. Give me words of wisdom to make men grow strong to face danger, wounding and even death itself, because their hearts are right with Yours. Help me to help them to live clean and to die brave, with no hatred in their hearts but only a love of what is right and a will to defend it. Grant me much grace to stand before them as another Christ, manly, true, fearless, and holy. O King of Peace, protect us while we fight — but grant us peace soon. Amen.

CLERICAL INDEX

Alter, Karl J., Most Rev., 136
Andruskevitch, Casimir, 152
Arnold, William R., Rt. Rev. Msgr., 152
Aubin, Jean Marie, S.M., Most Rev., 82, 161

Babst, Julius, 162
Ballinger, Francis, 35, 144
Barge, Henry, C.PP.S., 158
Barnes, B. L., xiii
Barr, Harold J., 105
Barry, Frank J., 158
Barry, Joseph, S.J., 178
Baumann, Herman C., 159
Bennett, William J., 157
Bergan, Gerald T., Most Rev., 139
Bonhomme, Joseph, Most Rev., 89
Bonn, John Louis, 156
Bouchey, Wilfred, 161
Brach, Stanley, 161
Bradley, Edward, 105
Brady, John J., 155
Brady, Terrence, 152
Braun, Albert, O.F.M., 159
Brody, Thomas V., 162
Buddy, Charles F., Most Rev., 115
Burns, John, 154
Busse, Julius S., 153

Cahill, Francis, 153
Callahan, John L., 156
Carberry, Richard E., 159
Carey, James A., 19
Casey, George, 52

Cavanaugh, John, 67
Clasby, William J., 25
Conerty, Thomas I., 154
Connelly, Richard J., 19
Conway, Anthony J., 161
Conway, J. D., 158
Cook, Ozias, B., 152
Couppe, Louis, M.S.C., 75
Crimmins, Harry B., 158
Culliton, John F., 56, 58, 91, 105
Cummings, William T., M.M., 159
Curran, John L., O.P., 156

Day, John B., 152
deKlerch, Emery, S.M., 86
Delahunt, Joseph B., 96
Deschenes, F. J., 158
Diamond, Martin J., 158
Dineen, Joseph, 152
Donnellan, Francis X., xiii
Donohoe, Hugh A., xiii
d'Orgeval, Peter, S.S.C.C., 165
Dorn, Frederick G., 165
Douaire, Arthur K., 141
Dougery, Edward, 138
Dougherty, Gerald M., O.S.M., 157
Doyle, Neil J., 31, 38 ff., 57, 121
Driscoll, Thomas P., 162
Drury, Thomas J., 50
Duehren, Joseph, 162
Duffy, John E., 159
Dunford, James, 160
Dugan, John J., S.J., 159
Dupuis, John M., C.S.C., 162

[181]

Duross, Thomas A., S.J., 91

Eagen, Richard J., 158
Early, Francis, 159
Early, Stephen B., S.J., 94
Egert, Clifford A., 158
Epalle, Bishop, 75

Fallon, Lester J., 94
Falter, Clement M., C.PP.S., 31, 35 ff.
Farren Edward J., S.J., xiii
Feisst, Cyril, 161
Finan, Arthur, C.SS.R., 151
Finnegan, Terrence P., 16, 82
Fitzgerald, James J., 87, 117, 155
Flanigen, George J., 105, 151
Floquet, Pierre Rene, S.J., 9
Flynn, James P., 31, 41 ff.
Flynn, Patrick J., xii, 1
Foley, John E., 155
Ford, Henry F., 66
Forest, J., O.F.M., xiii
Foster, Thomas E., 49
Fournie, John T., 161
Freeman, Justin E., O.S.B., 153

Gaffney, F., 90, 98
Gales, Louis A., xiii
Gallagher, John, C.SS.R., 142
Gannon, Bernard, 179
Gearhard, August F., 79, 147
Gehring, Frederic P., 87, 118
Geigel, Frank, 159
Gilloegly, James, 154
Gorman, Francis, 88, 154
Grassi, Bishop, 77
Guenette, Alfred J., 18

Haddigan, Michael, 153
Haggerty, James J., 152
Hawko, Richard J., xii
Hayes, William J., O.F.M., 172
Healy, Bernard, xiii
Hearn, Robert, C.SS.R., 151
Heavey, William J., S.J., 138

Heindel, Elmer, 161
Henry, Austin, 158
Hoffmann, Albert J., 63
Hrdlicka, Joseph, C.SS.R., 151
Hubbard, Bernard, S.J., 160
Hurley, Roderick, O.Carm., 162

Kane, Stephen W., 133
Kearney, James E., Most Rev., 161
Keefe, Anselm M., O.Praem, 157
Keenan, Francis J., C.M., 154
Kelleher, Jeremiah, 88
Kelly, Francis W., 157
Kelly, John P., 33
Kempker, George M., 152
Kennedy, Hugh F., S.J., 159
Keough, Matthew F., 155
Kerwin, Paschal E., O.F.M., 175
Kilfoil, J. Hugh, 162
Knier, Albert, xiii
Kusman, Stanley J., S.M., 155

LaFleur, Joseph V., 156
Landry, John N., 162
Lane, James C., 173
Leonard, John E., 99, 105
LeSage, Leo, 161
Lew, Edward L., 158
Lippert, Paul, 24
Liston, James M., 18, 151
Lollich, F. J., 158
Lotbiniere, Francois Louis Chartier, 9
Loughlin, John V., 161
Lynch, Edward, O.M.I., 34
Lyons, Michael J., 157

Madden, Leo W., xii, 47
Maguire, William A., 155
Mahoney, John P., 57
Maloney, John S., 132
Mannion, Harry P., 161
Mannion, Joseph Patrick, 105
Martin, Kenneth C., 158
Matthews, (?), 170

McAuliffe, John, 158
McDonald, L. J., 152
McDonnell, John J., 159
McDonnell, Thomas J., Rt. Rev. Msgr., xii
McElroy, John, S.J., 10, 12
McEvoy, James E., xii
McGarrity, John Joseph, 155
McGinnis, James Sidley, S.J., 157
McGoldrick, Joseph, 138
McGuire, John P., 56, 103
McManus, Francis J., 156
McNulty, Peter E., xii
McQuaid, Arthur C., 55
Meany, Stephen J., S.J., 155
Meehan, Daniel F., 80
Meighan, Matthew, C.SS.R., 143
Menster, William J., 156
Michaels, Emmett T., 162
Molyeux, Robert, 9
Moore, Denis G., 161
Moorman, E. A., C.PP.S., 158
Morczka, John A., 50
Mulhern, Philip F., O.P., xii
Mulligan, Regis J., C.P., 50
Murphy, Francis X., 153
Murphy, John P., 154, 159
Murphy, Joseph B., C.S.Sp., 17

Nebel, Charles A., 162
Newman, Philip J., 152
Niglis, Joseph, 159
Norkett, Edward J., 50

O'Brien, Daniel, C.SS.R., 151
O'Brien, James W., 156
O'Brien, Walter, O.F.M.Cap., 152
Obrist, Lawrence F., 159
O'Connor, John A., xii
O'Connor, William T., 158
O'Donnell, James J., 156
O'Gara, Harold P., 157
O'Keefe, Eugene J., S.J., 159
O'Leary, Francis T., 35
O'Malley, Thomas, 153
O'Neill, Ralph M., S.J., 99

O'Neill, William R., 145
Owens, Joseph P., 162

Parks, Charles H., 155
Parsons, Wilfred, S.J., 74
Patrick, C. A., xii
Paulekas, Alphonsus J., C.SS.R., 177
Pixley, William, C.S.Sp., 17
Power, Richard D., 155
Powers, John, C.SS.R., 105, 144

Quest, Charles, 157
Quinleven, George T., 161

Ready, Michael J., Rt. Rev., 18
Reardon, Thomas M., 73, 87, 152
Redmond, Paul J., O.P., 152
Reedy, John J., 98
Regan, Thomas F., 155
Reilly, Francis, 157
Reilly, Stanley J., 159
Rey, Anthony, S.J., 10, 12
Rice, James, C.SS.R., 144
Rigney, Harold W., 89
Ronan, Edwin, C.P., 158
Ruhl, E. H., 158
Ryan, Daniel J., 162
Ryan, Patrick J., 35, 160

St. John, John D., S.J., 173
Scannell, John W., 153
Scecina, Thomas, 159
Schmieder, Lawrence R., 162
Schmitt, Aloysius Herman, 31, 33 f.
Schomer, B. V., 154
Schultz, John G., C.SS.R., 151
Schwartz, Aloysius, 152
Schwener, Henry, 152
Scully, Richard, 154, 162
Seeman, L. W., xiii
Shannon, Francis X., 154
Sherwood, Gervase, 157
Siwinski, Clement A., 158
Slattery, Edward A., 170

Slavin, William M., 154
Sliney, Edmund C., 74
Spellman, Francis J., Most Rev., 65
Stack, Kenneth G., 95, 115
Steffens, Albert, 157
Stober, Henry B., 159
Strub, Herman, 158
Sullivan, Ambrose, 156
Sullivan, Francis V., 154

Talbot, Albert D., S.S., 159
Targosz, Stanislaus, 102
Tennessen, M. M., 139
Thompson, Ralph A., 158
Tikuisis, Vincent, 162
Tooher, Thomas, 154
Toomey, J. E., 158
Toomey, Patrick J., C.S.V., 158
Torralba, Aloysius F., S.J., xii

Valenberg, Bishop, 74
VanBeersum, Anthony G., 151
Vanderheiden, Joseph, O.S.B., 159
VanGarsse, William, 162
Verius, Stanislaus, M.S.C., 76

Vesters, Gerard, Most Rev., 75

Wade, Harry F., C.SS.R., 145
Wagner, Sylvester, 154
Wall, John J., 87, 157
Walsh, Joseph J., 132
Walsh, Vincent, 154
Waters, Edward J., 134, 156
Wheaton, John K., 61
Whelan, Gerald, C.SS.R., 151
Whelan, Joseph E., 149
Wieber, Joseph E., 147
Wiley, S. Ernest, 157
Wiley, Robert, 157
Wilson, John, C.PP.S., 36
Wise, John S., C.SS.R., 151
Wissert, John P., C.PP.S., 158
Wolfe, Thomas J., 154
Wollack (Woloch), Francis, 169
Wood, John J., S.C.V., 157
Wurm, Urban J., 137

Yeager, C. J., 154

Zellens, Aubrey, O.S.B., 157
Zerfas, Matthias E., 159
Ziober, Anthony A., 157

Date Due

DEC 6 '44			
DEC 1 8 '44			
JAN 4 '45			
JAN 1 8 '45			
FEB 6 - '45			
MAR 1 5 '45			
APR 3 - '45			
APR 4 - '45			
MAY 1 4 '45			
JUL 1 8 '45			
N 1 9 '46			
8 2 0 '47			
JAN 1 9 '50			
MAR 1 0 '54			
DEC 1 8 '61			

HUNTTING—1320

MARYGROVE COLLEGE LIBRARY
War is my parish,
940.5478 G76

3 1927 00128075 6

940.5478
G76   Grant, Dorothy (F.)

    War is my parish

DATE	ISSUED TO
	Anne Youngblood

940.5478
G76